# The Invisible Universe
## SECOND EDITION

Gerrit Verschuur

# The Invisible Universe
## The Story of Radio Astronomy

SECOND EDITION

 Springer

Gerrit Verschuur
University of Memphis
Memphis, TN 38002

ISBN   978-1-4419-2156-7          e-ISBN   978-0-387-68360-7

Printed on acid-free paper

Printed on acid-free paper.

9 8 7 6 5 4 3 2 1

springer.com

*Dedicated to the memory of Heinrich and Elisabeth Hertz who lit the spark that illuminated the invisible universe.*

# Acknowledgments

When the first version of this book was published back in 1973 it was possible to summarize all of radio astronomical discoveries in fair detail in a single monograph without overwhelming the reader. That was because the science of radio astronomy was barely 20 years old since technology had spurred a rapid growth in our ability to map the heavens in the radio band. The subsequent edition of this book, published in 1987, entitled The Invisible Universe Revealed, reflected the rapid growth of this science by including dramatic radio images, or radiographs, of distant sources of radio waves.

At the start of the 21st century our ability to produce stunning images of radio galaxies, for example, is a matter of routine and color is readily added for effect. In this edition we have included several of the most informative colorized radiographs made to date. Another enormous change seen over the last 20 years is the sheer volume of information that has been accumulated by a generation of very large radio telescopes working over an increased wavelength range. Therefore it is no longer possible to provide a comprehensive overview such as the one that made up the 1973 edition.

It is with this caution in mind that we enter the Invisible Universe of radio astronomy to describe its contents in broad terms, cognizant that to go into more detail would make this book unacceptably long (and expensive!). Also, in producing this new edition I have kept in mind the interests and potential needs of the intelligent lay person who might have visited a radio observatory and who then seeks to assuage their curiosity by reading more about this science.

The present rewrite would have been impossible without the input of a large number of colleagues, some in person, others through email, and the help of staff at the National Radio Astronomy Observatory. Also, each of the images shown in this book required a large amount of careful work on the part of dozens upon dozens of colleagues who together have turned the invisible universe of radio astronomy into a gallery of visual representations that stagger the imagination. Thus it is with great pleasure that I list those who have helped in re-educating me, and those who provided information that made my task possible. I apologize for any inadvertent omissions.

Thank you Sue-Ann Heatherly, for encouraging me to take on the project, and my wife Joan Schmelz for seconding the motion and for her subsequent urging and encouragement as well as her editorial advice, to Barry Turner for having been a great office mate for years in the youthful days of radio astronomy, and for helping establish a reliable molecule list, with assistance from Al Wooten, and Butler Burton for his overall enthusiasm for the project and help in getting it started.

This revision would not have been possible without the help of the following colleagues: Fred Lo, Miller Goss, Patricia Reich, Rainer Beck, Cornelia Lang, John Hibbard, Dave Hogg, Scott Ransom, Jim Moran, Mark Reid, Harvey Lizst, Peter Kalberla, Ed Fomalont, Ken Kellermann, Alan Bridle, Katherine Blundell, Meg Urry, Juan Uson, Alan Bridle, Elly Berkhuijzen, Dave Jauncey, Charles Lada, Tom Dame, Jim Braatz, Steve Schneider, Baerbel Koribalski, Lister Staveley-Smith, Paul Vanden Bout and to NRAO staff members Billie Rodriguez, Pat Smiley, and Marsha Bishop.

Thank you one and all, and enjoy.

# Contents

# Introduction: Adventure, Imagination, and Curiosity

## The Exploration of the Radio Astronomical Unknown

Radio astronomy is one of the great adventures of the human spirit. Exploratory behavior, the primal urge that drives us into the unknown, is rooted in curiosity and expressed in a deep human hunger for venturing into new worlds, a hunger that has been dramatically expressed in thousands of years of slow, systematic, and sometimes frightening journeys of exploration and evolution. Such journeys, overland and across the seas and oceans, have carried people from their birthplaces to the most distant corners of the planet and farther. Like pollen on the wind, our species has moved from the caves of earth to the craters of the moon. Our instinct drives us on, not just to the planets, but further, into the universe beyond our senses where profound mysteries have been uncovered, mysteries that challenge our imagination and our capacity for comprehension.

Radio waves from space carry information about some of the most intriguing natural phenomena yet discovered by human beings. This is the bailiwick of radio astronomy. However, the cosmic radio whispers reaching the earth compete with the electrical din produced by TV, radio, FM, radar, satellite, and cell phone signals. Thus the faint radio signals from space that memorialize the death of stars, or tell of awesome explosions triggered by black holes in galaxies well beyond sight, are nearly lost against the background of human-made static. Yet such radio waves contain the secrets of interstellar gas clouds and carry messages from the remnants of the Big Bang that propelled our universe into existence.

In order to gather the faint cosmic signals and avoid the unwanted stuff, astronomers use powerful radio telescopes located far from cities. Those telescopes are huge metal reflectors that focus the electromagnetic messages from space, which are then amplified in sensitive receivers and fed to computers where they are converted into a visual form to be displayed, analyzed, interpreted, and hopefully understood.

The story of radio astronomy is a tale of the constant quest to express in clearer visual forms the information carried by the radio waves. For this reason radio astronomers are always inventing new techniques to allow them to "see" the radio

1

sources more clearly. The better we "see" the sources of those radio waves, the more likely we may be to understand their inner secrets.

Ever since Galileo first turned a telescope toward the heavens in 1609 AD, centuries of technological innovation have afforded an increasingly clear view of astronomical objects in the far reaches of space. Larger and more sophisticated telescopes are always being designed and constructed. Today, modern technological marvels such as the Hubble Space Telescope, the mightiest optical telescope ever built, allow astronomers to perceive the visible universe with fabulous clarity. Not to be outdone, giant radio telescopes reveal the radio universe in comparable detail, and they have opened our imagination to a cosmos beyond our senses in a way previously undreamed of.

## Seeking New Knowledge

Like any science that seeks answers beyond the borders of the unknown, radio astronomy requires a great deal of thought and effort and, especially recently, significant amounts of money. In asking governments for funds to construct a new radio telescope, the modern explorers of space are following a time-honored tradition. Voyages of discovery have always been costly affairs, usually sponsored by monarchs, business interests, or empires. Even Columbus needed a "research grant" from Qeen Isabella to carry him across the ocean. Today, tax dollars fund scientific instruments, the new vessels of discovery, and the scientist/explorer's challenge has become far more subtle than it once was.

In ancient times the sponsor of an explorer's journey had an expectation that the ship would return with a cargo of spices, gold, or silver—something that could be used in barter. It is no longer so. The new explorer searches for knowledge–subtle, ethereal knowledge. This may be returned in the form of a radio image of a distant galaxy or of the invisible center of an interstellar gas cloud. It is impossible to attach financial worth to such images, just as it is impossible to attach value to any bits of that elusive substance called knowledge. What is clear, however, is that many of the pictures of radio sources in this book are beautiful in their own right even as they reveal the existence of previously unknown phenomena, knowledge of which broadens our perspectives about the universe into which we are born.

## Radio Astronomy and Imagination

The invisible universe of radio astronomy is revealed in images that startle the imagination. Although this book contains only static pictures, each radiograph is a snapshot of an object in a state of continual upheaval. The motion, the chaos, and the violence found in the invisible universe can only be recognized when you wrap your imagination around the images. Do not hesitate, because your imagination is as valid as the next person's in trying to visualize this.

Full appreciation of the new discoveries requires the continual involvement of your imagination. Exploration of the cosmos becomes an adventure when it takes place in the mind. The explosion of a quasar is not witnessed in space somewhere, but in your imagination. All we see out there, now, is but an instantaneous snapshot of what took place millions or even billions of years ago.

The dynamical aspects of astronomy are revealed not by what is seen at the far end of the telescope, but what is experienced at this end. This is where the excitement is to be found. Thanks to the workings of the human mind, aided by physics, mathematics, and computers, astronomers can simulate cosmic phenomena that allow us to recognize how evolution, change, and catastrophic events shape distant gas clouds, dying stars, galaxies, and quasars.

The human race looks out into space and discovers marvelous beauty, a beauty that often lies beyond our normal powers of perception. Yet it is a beauty that can touch us as profoundly as any terrestrial sunset, symphony, or songbird. In radio astronomy the beauty is perceived by fully harnessing our imagination as we travel beyond the senses. In the following pages you will join in the adventure and share the excitement of exploration as we journey into the invisible universe, a universe revealed by tiny amounts of radio energy reaching us from millions or even billions of light-years away.

# 1
# What is Radio Astronomy?

## 1.1. A Little History

In 1886, Heinrich Hertz accidentally constructed the first radio transmitter and receiver. In a darkened lecture theater at the Technical College in Karlsruhe, in Germany, Hertz had set up an experiment to test what happened when an electrical current flowed in an open circuit (that is, a circuit with a gap in it). As he explained the setup to his wife, Elisabeth, he switched on a spark generator, used to produce current, and one of them noticed a simultaneous spark that flashed in an unrelated piece of equipment at some distance away from his main experimental apparatus. Whoever noticed it first, Heinrich or Elisabeth, is unknown to us, but it was Heinrich who made the leap of curiosities that underscore the nature of scientific research. Hertz asked "Why?" and started a systematic search for an answer.

Eighty years later historians of science would report that Hertz was at least the sixth physicist to see this odd effect, but he was the first to follow up on his key question. He proceeded to design a series of brilliantly simple experiments, one after another, in search of an answer. He was able to show that an invisible form of radiation, which he called "electric waves," carried energy through intervening space. Hertz was also able to demonstrate that the electric waves were a phenomenon very similar to light. In fact their speed through the air was the same as that of light. Today we know that both light and Hertz's "electric waves" are forms of electromagnetic radiation (see Appendix A.2). Over time, the Hertzian waves (a name used very early in the 20th century) came to be called radio waves. Their frequency is measured in cycles per second, now called Hertz (Hz). In Appendix 2.1 the relationship between frequency and wavelength is discussed. For the bulk of our story we will refer to the frequency of radio waves.

Hertz died tragically at the young age of 35 of blood poisoning from an infected tooth. If he hadn't, he surely would have won a Nobel Prize in Physics for his discovery.

After showing that radio waves behave much as light does, except that they are utterly invisible, Hertz did not ask how far they might travel through space. That was left to Guglielmo Marconi, the Italian physicist who performed a series of obsessively creative experiments to prove that radio waves could travel enormous distances and even pass through rock. He was wrong in this latter belief, but he did show that a radio signal could traverse the Atlantic Ocean. The reason that the radio waves made it across despite the curvature of the earth was because the earth's atmosphere is surrounded by an electrically conductive layer known as the ionosphere and radio waves bounce off that layer to be reflected across the ocean. That wouldn't be understood until decades later. Meanwhile, Marconi was happy to know that radio waves did go all the way around the earth and it was not long before that ships at sea could signal one another and to their home ports using radio waves. By 1912 the infamous sinking of the Titanic spread awareness that radio transmitters could send an SOS far and wide.

Marconi did wonder whether there might be radio waves reaching earth from space but his equipment would not reveal the existence of the wondrous invisible universe for the same reason that he could signal across the Atlantic. At the low radio frequencies that Marconi used, the reflecting ionosphere not only allows radio signals to bounce around the curvature of the earth, it also prevents radio waves from space from reaching the earth's surface. Those that do arrive from space are reflected back. (Only if their intrinsic frequency is higher than about 20 MHz such radio waves do reach the ground unimpeded, but then it was not known very much about building receivers at such frequencies.)

## 1.2. The Birth of Radio Astronomy

Karl Guthe Jansky, the father of radio astronomy, was employed at Bell Laboratories, which, in 1927, introduced the first transatlantic radiotelephone. For a mere $75 one could speak for three minutes between New York and London, but the radio links were terribly susceptible to electrical interference. The first system operated at the extraordinarily low frequency of 60 kHz (that is, at the very long wavelength of 5 km) and in 1929 a change was made to short waves whose frequencies were in the range of 10–20 MHz. But the new telephone links were still susceptible to electrical disturbances of unknown nature, which plagued the connections. Jansky was assigned the task of locating the source of the interference. To carry out his studies he built a rotating antenna (Figure 1.1) operating at 20.5 MHz and by 1930 began taking regular observations. In 1932 he reported that local and distant thunderstorms were two sources of the radio noise and a third source was "a very steady hiss-type static, the origin of which is not yet known."

During the following year he was unambiguously able to demonstrate that the source of the signals was outside the earth and presented a report entitled "electrical disturbances apparently of extraterrestrial origin." And so radio astronomy was born.

Just imagine this: When Jansky became convinced he had picked up radio waves from space he enjoyed what few people ever experience—the thrill of discovery—seeing something no one had ever seen before. That is part of the reward, the joy, and excitement of doing scientific research.

FIGURE 1.1. Karl Jansky, working at Bell Telephone Laboratories in Holmdel, NJ, in 1928 built this antenna to receive radio waves at a frequency of 20.5 MHz (wavelength about 14.5 meters). It was mounted on a turntable that allowed it to rotate in any direction, earning it the name "Jansky's merry-go-round." By rotating the antenna, he could find what the direction from which radio signals were arriving. In this way he identified radio waves coming from the Milky Way, in particular its center inn Sagittarius. (Image courtesy of NRAO/AUI.)

Fifty years later, at the National Radio Astronomy Observatory in Green Bank, West Virginia, distinguished radio astronomers gathered to celebrate the anniversary of Jansky's discovery. A report entitled "serendipitous discoveries in radio astronomy" grew out of that meeting and it presents the human side of the birth and growth of this science.

"Serendipity" is a term coined by Horace Walpole, the writer and historian, who used it to refer to the experience of making fortunate and unexpected discoveries, according to the fairy tale about the three princes of Serendip (an old name for Ceylon). Serendipitous discoveries are those made by accident, but also by wisdom; however no one can make an accidental discovery unless that person is capable of recognizing that something of significance is occurring. Jansky was such a person.

In January 1934, in a letter to his father, Jansky wrote:

Have I told you that I now have what I think is definite proof that the waves come from the Milky Way? However, I'm not working on the interstellar waves anymore.

His boss had set him to work on matters of more immediate concern, matters, which were:

...not near as interesting as interstellar waves, nor will it bring near as much publicity. I'm going to do a little theoretical research of my own at home on the interstellar waves, however.

Jansky did not take interest in his new discoveries to the point of building his own antenna so as to pursue his explorations over the weekends. Jansky's boss, who ruled with an iron hand, was later to encourage him to write another report, and in 1935 Jansky interpreted the sky waves as coming from the entire Milky Way. But he did not know why and suggested that either a lot of stars were contributing or something in interstellar space was the cause. He realized that if the waves were due to stars he should have picked up the sun. As observed from the surface of the earth the Milky Way happens to reach its maximum brightness in the radio band close to Jansky's chosen frequency. It is brighter at still lower frequencies, but those radio waves do not penetrate the ionosphere. Furthermore, the ionosphere experiences daily changes of its characteristics and in the daytime blocks out the sun's radio emission. Thus Jansky's antenna was blind to its radiation. The mid-1930s were also a time of sunspot minimum, which meant that the ionosphere was transparent to 20 MHz at night. Had Jansky been observing at sunspot maximum, the ionosphere, whose reflecting properties vary with time of day and season as well as sunspot cycle, would have blocked out all 20-MHz radio waves from space and he would not have discovered the signals from the Milky Way.

Jansky failed to pursue his discoveries any further because there were other projects to be done and "star noise could come later" he was told by his employers. It was to be years before significant follow-up work began. A few astronomers in the United States and Europe had become aware of Jansky's work, but any plans to look more closely at his discoveries had to be shelved when World War II broke out. In any event, most astronomers knew absolutely nothing about radio receivers and antennas, so how could they get involved? Optical astronomers were just not equipped with the skills necessary to tinker with radio sets. After the war it was mostly radio physicists who launched the new science, and they had to learn astronomy in the process.

After he made his discovery, Jansky wrote to the famous physicist, Sir Edward Appleton:

If there is any credit due to me, it is probably for a stubborn curiosity that demanded an explanation for the unknown interference and led me to the long series of recordings necessary for the determination of the actual direction of arrival.

Such stubborn curiosity is the hallmark of good scientists. Jansky trusted his data and continued his measurements for confirmation. His persistence led to the discovery that the source of the static lay in the astronomical heavens.

The story of radio astronomy is replete with apparently amazing cases of fortuitous discoveries, but such discoveries requires more than good luck. They require a prepared mind and dedicated effort to follow up on what might at first have seemed to be a preposterous new observation. What would have been more preposterous, at least back in 1933, than to learn that radio waves were reaching the earth from all sorts of strange astronomical objects, and even from the beginnings of time and space.

## 1.3. So What Is Radio Astronomy?

Radio astronomy involves the study of radio waves from the depths of space. Many objects in the universe, including stars, galaxies, and nebulae, as well as a wide variety of peculiar, fascinating, and often mysterious objects, emit radio waves through naturally occurring processes.

### *1.3.1. How Radio Waves from Space are Generated*

Cosmic radio waves are created in several ways, depending on the physical conditions in the radio-emitting object. All the processes involve the movement of electrons, in particular, changes in their velocity during which the electrons lose energy, which can be radiated away as a radio wave. Radio energy is produced either by slow-moving electrons (traveling between tens and hundreds of kilometers per second) within hot clouds of gas that surround very hot stars, for example, or by electrons that have been accelerated to near the speed of light through stellar or larger scale explosions, which energize the particles. The two radio emission processes are known, respectively, as thermal and nonthermal.

Nonthermal emission, sometimes called synchrotron radiation, involves cosmic ray electrons that spiral around magnetic fields and radiate energy in the form of radio waves. Depending on the energy of the particle and the strength of the magnetic field involved, this process can produce emission at any of the wavelengths across the electromagnetic spectrum (see Appendix A.2).

### *1.3.2. Radio Telescopes*

For hundreds of years, ever since Galileo, in 1609 AD, first used an optical telescope to study the moon, stars, and planets, astronomers have used glass lenses or a mirror to gather and concentrate light from distant stars and galaxies. The light is then passed through more lenses to bring it to focus on a photographic plate or on an electronic detector in the best of modern telescopes.

A radio telescope is similar to an optical telescope, but it reflects radio waves off a metal surface instead of a glass mirror. The larger the reflecting surface the greater the amount of energy gathered and the fainter the radio signals that can be sensed. For decades the 250-ft diameter Lovell telescope at the Nuffield Radio Astronomy Laboratories of the University of Manchester in England was the largest fully steerable radio telescope in the world, see Figure 1.2.

FIGURE 1.2. The 250-ft diameter radio telescope at the Jodrell Bank Observatory in England. At the top of the 60 ft high mast in the center of the dish a box houses electronic amplifiers attached to a small antenna mounted just beneath the box at the focus of the telescope. The amplified signal is then fed down to a laboratory at ground level or, in the days that I was using the dish, to our lab in the tower at the right. This image is of the upgraded version known as the Lovell telescope. The version we used in the early 1960s had only one stabilizing girder carrying much of the weight and the surface was not as carefully constructed. (Photo courtesy of Nuffield Radio Astronomy Laboratories, University of Manchester.)

In January 1961 I arrived at Jodrell Bank to begin my postgraduate work, having just arrived from South Africa. It was a foggy day and I could not see the giant telescope until I was nearly under it. Then I was overwhelmed by its awesome size as it loomed through the fog. That image remains deeply etched in my memory. Our lab was situated 120-ft above ground in the enclosed space seen in Figure 1.2 in what was called the green tower, on the right. The walls of this lab were made of steel plates and had no insulation. The important elements of the receiving system were in a small, heated enclosure that barely allowed us to enter to make adjustments when needed. During the infamous winter of 1962–1963 in England the temperature inside this lab dipped below freezing and remained there for weeks despite having heaters going to keep us warm. After tedious trips up to the focus located 60 feet above the dish, which required a 4-minute one-way ride in a small funicular platform, any attempts to warm my feet by an electric fire upon my return to the metal box tended to set my socks smoldering before I even knew that the heat was on. Our midnight observing runs, which required remaining awake from midnight until 10 am tending the paper charts recorders, were a test of endurance. They were also dramatic and fun.

Radio waves from space are reflected off the parabolic surface to a focus (top of the mast in the center of the dish in Figure 1.2) where a small antenna is placed that may look similar to a conventional TV or FM antenna but far more. There the concentrated radio signals that are converted into minute electrical currents in amplifiers connected to the antenna. This is known as the "front end" of the receiver. These currents are then sent to the control room where they are amplified a million or more times in the "back end" of the receiver before being processed in a computer or displayed in such a way that the radio astronomer can "see" what the data indicate.

A single-dish radio telescope will collect all the radio energy coming from some small area in the heavens at any instant. That area is called the beam and defines the resolution of the telescope, which depends on the observing frequency and the diameter of the dish. The larger the diameter or the higher the frequency, the better the resolution (smaller beam width). The 250-ft radio telescope shown in Figure 1.2 has a beamwidth of about 12 arcminutes at a frequency of 1400 MHz.

In order to produce the equivalent of a photograph of a section of sky the radio telescope has to be systematically "scanned" in the same way that a TV image is produced by scanning an electron beam across the TV screen. The intensity of received radio signals is recorded and the data combined to produce a radiograph, the visual image of what a particular direction in the sky looks like to the radio telescope.

## 1.3.3. What is a Radio Source?

One of the earliest post-World War II discoveries in radio astronomy was that specific regions of the sky seemed to emit more radio energy than their surroundings. These were given the generic name of "radio source." Whenever a larger radio telescope or more sensitive radio receiver was used, more radio sources were discovered. Today tens to thousands of radio sources are known.

The accuracy with which the first radio sources were located in the sky was insufficient to allow optical astronomers to decide which of the hundreds or thousands of images of stars, galaxies, and nebulae in their photographs of the region in question was responsible for the radio emission. In order to make an optical identification the astronomers required an accuracy of 1 arcminute or less (Appendix A.6), although by the late 1940s and early 1950s half a dozen of the strongest radio sources had been identified with obviously unusual, and hence interesting, optical objects. Those included a couple of nebulae associated with the remains of exploded stars, and several distant galaxies.

The list of various types of radio sources now known includes stars, nebulae, galaxies, quasars, pulsars, the sun, the planets, as well as amazing clouds of molecules between the stars, all of which generate radio waves. The study of the cosmic radio waves—where they come from, how they are produced, what sorts of astronomical objects are involved—is what radio astronomy is all about.

# 1.4. Radio Interferometers

In order to "see" more clearly the radio astronomer needs, above all, high resolution. As stated above, the larger the diameter of a single dish antenna the better its resolution, but there is a limit to how large a structure can be built before it collapses under its own weight. Instead, radio astronomers began to combine the signals from two dishes separated by miles in what is called an interferometer. Its resolution is set by the maximum distance between the component dishes.

A very beautiful variation on this was developed at Cambridge by the radio astronomers led by Sir Martin Ryle. In this technique the aperture (or area) of a very large dish is synthesized by many small dishes set far apart, and feeding their individual signals to a powerful central computer.

Imagine two 10-meter diameter dishes located on a football field and pointed at a given radio source. If you store the radio signals from each of these dishes as they are moved to every point on the field and then combine all the data, it is possible to synthesize what you would have observed had you used a single dish of the size of the entire football field. What Ryle and his team realized was that, as seen from the radio source, any two radio telescopes appear to move around each other during the day due to the rotation of the earth. That means you don't have to physically move the dishes. You just let the earth do the walking. Enormous apertures can be synthesized in this way.

In practice, aperture synthesis involves using an array of dishes spread over dozens of miles of countryside.

## 1.4.1. Very Large Array

The world's largest aperture synthesis telescope is the Very Large Array (VLA), 50 miles west of Socorro, in New Mexico, one of the National Radio Astronomy Observatory NRAO's repertoire of beautiful radio telescopes (Figure 1.3). Observations with the VLA were used to make many of the radiographs shown in this book. Twenty-seven individual radio antennas of 25 meter diameter are located along railroad tracks, which are laid out in a Y-shape, each arm of which is 21 km long. To completely synthesize the largest possible aperture obtainable by the VLA the individual antennas have to be moved to different locations along the rail tracks every few months.

One of the most stunning images made by the VLA is of the radio source known as Cygnus A, shown in Figure 1.4.

## 1.4.2. Very Long Baseline Array

The Very Long Baseline Array (VLBA) is a continent-sized radio telescope (Figure 1.5) is capable of enormously high resolution. Ten antennas are located from St. Croix in the Virgin Islands, to Hawaii, with eight distributed over the continental United States. As with all new antenna arrays, the resulting radio telescope

FIGURE 1.3. The VLA radio telescope of the National Radio Astronomy Observatory located west of Socorro, NM, out in the middle of nowhere, which is the way radio astronomers like it. In this view many of the dishes are spaced in the so-called compact array. The railroad tracks on which the dishes can be moved for up to 11 miles along each of three arms of the array can be seen in the foreground. (Image courtesy of NRAO/AUI.)

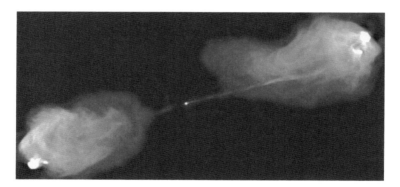

FIGURE 1.4. A radio image, or radiograph, of Cygnus A, one of the most powerful sources of radio waves in the heavens, as observed with the VLA. Tenuous filaments of radio emitting gas constrained by magnetic fields illuminate two enormous lobes fed by jets blasting out of on either side of a central galaxy located 600 million light years distant (see Chapter 10). Investigators: R. Perley, C. Carilli, and J. Dreher. (Image courtesy of NRAO/AUI.)

FIGURE 1.5. The location of the 10 dishes that make up the VLBA radio telescope of the National Radio Astronomy Observatory. Data from all the out stations are brought together at the central processing computer in Socorro, NM. (Image courtesy of NRAO/AUI.)

operates on aperture synthesis principles. The VLBA can attain an angular resolution of two tenths of one thousandth of an arcsecond (0.2 milliarcseconds), which may be compared with 1 arcsecond for the typical radiographs shown in this book.

# 2
# A Science is Born

## 2.1. Caught between Two Disciplines

In 1933 John Kraus, then at the University of Michigan, attempted to detect the sun by using a searchlight reflector to focus the radio waves. He failed because the receiver was not sensitive enough. This was the first use of a reflector-type radio telescope. At the Serendipity meeting, Kraus stated that meaningful accidental discovery occurs only as the result of "being in the right place with the right equipment doing the right experiment at the right time." Another noted astronomer, R. Hanbury Brown, added that the person should "not know too much," otherwise the discovery might not be made!

This summarizes a very interesting phenomenon. Many research scientists, especially the theoretically inclined, "know" so much that their chance of making a lucky or creative discovery may be severely curtailed. If we know too much, our vision is sometimes narrowed to the point where new opportunities are not seen. Jansky knew a little astronomy, but not enough for it to get in his way and cause him to reject the possibility that radio waves originating in the cosmos might be real.

Grote Reber, a professional engineer and radio ham in his spare time, was one of the few people who recognized the interesting implications of Jansky's discovery. Reber was certainly not hampered by any astronomical prejudices about whether or not the cosmic radio waves could exist. Instead, he was interested in verifying their existence and followed up on Jansky's work. To this end, Reber built the world's first steerable radio dish antenna (Figure 2.1) in his backyard and mapped the Milky Way radiation during the period 1935–1941. Figure 2.2 shows an example of Reber's data. He pointed out that the new field of radio astronomy was originally caught between two disciplines. Radio engineers didn't care where the radio waves came from, and the astronomers

... could not dream up any rational way by which the radio waves could be generated, and since they didn't know of a process, the whole affair was (considered by them) at best a mistake and at worst a hoax.

The very essence of research is that once an observation is made it requires some understanding and interpretation in order to formulate a plan for making further

FIGURE 2.1. Grote Reber's telescope in his back yard in Wheaton, Illinois, about 1938. (Image courtesy of NRAO/AUI.)

observations. It was initially very difficult for astronomers, entirely ignorant of radio technology, to interpret or understand the significance of Jansky's or Reber's epoch-making discoveries.

Jesse Greenstein, of Caltech, one of the few astronomers who did get involved before World War II, summed up the dilemma confronting the astronomer of those prewar days:

I did not say that the radio astronomy signals would go away someday, but I didn't know what next to do.

How could anyone know what next to do? The mystery of where the radio waves originated was a profound one, not easily solved. Significant new technologies

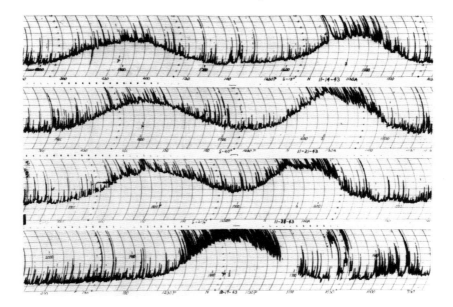

FIGURE 2.2. Chart recordings from Reber's telescope made in 1943. The spikes are due to terrestrial interference seen against the changing signal due to radio emission from the Milky Way as the radio telescope is scanned across the sky. (Image courtesy of NRAO/AUI.)

had to be combined with astronomical knowledge in order to carry out radio astronomical research. If the science was to flourish, either astronomers had to learn about radio engineering or radio engineers had to learn astronomy. The new science therefore grew slowly. The intrusion of World War II may have speeded up its growth because of the intense research in radar techniques, which led to the very rapid development of precisely those types of radio antennas and receivers that the radio astronomers were to require for their work. After the war those dishes and receivers became freely available as war surplus equipment.

## 2.2.  Postwar Years—Radar Everywhere

England, Australia, France, the Netherlands, the United States, and Canada were the important centers for postwar radio astronomy. The radio engineers and physicists drawn into radar research during the war became the first generation of professional radio astronomers. The equipment they used to launch their research work was scrounged, begged, or borrowed from military surplus.

In 1946 in Canada, Arthur Covington, of the National Research Laboratories, began taking regular observations of the sun at 3,000 MHz (10.7-cm wavelength), a choice dictated by the availability of surplus radar components. For decades this work was to provide the data for anyone interested in knowing how active the sun was on any given day. The solar radio data showed that the sun's radio brightness is directly correlated with the 11-year sunspot cycle and also revealed that the radio emitting regions on the sun must be at temperatures of over one million degrees.

Radio astronomers in the Netherlands did their early work with a German war surplus radar (Wurzburg) dish. Since 1944, when H. C. van de Hulst, a graduate student working with J. H. Oort, had given a talk on how radio observations might contribute to our understanding of the universe, the Dutch had focused their attention on searching for radio emission from hydrogen gas between the stars, with ultimate success in 1951 (see Chapter 6).

The Cambridge radio astronomy effort, under Martin Ryle, made heavy use of two Wurzburg dishes combined as an interferometer, which were used to accurately locate some of the strongest radio sources in the sky with sufficient accuracy that optical identifications could be made. In 1948 Ryle and F. G. Smith discovered a very strong radio source in the constellation of Cassiopeia, which came to be known as Cas A. This naming scheme reflected a naïve expectation that radio sources could be labeled by the constellation in which they were found and that using letters of the alphabet to indicate successively weaker sources would suffice. That system did not survive long and today tens of thousands of radio sources have been detected. Nevertheless, this appellation continues to be used for some of the first, and brightest, radio sources discovered back then.

Smith succeeded in improving the radio measurements to the point where an accuracy of 1 arcminute allowed the optical astronomers on Palomar mountain to photograph the position and discover the filamentary remains of a supernova in Cassiopeia that coincided with the radio source Cas A (see Chapter 4). The position of Cygnus A (Figure 1.4) was also measured accurately enough to lead to its identification with a very faint, distant galaxy.

During the later phases of the war, radar antennas in south England had been pointed just above the horizon to detect incoming V2 rockets, and in the process they accidentally picked up echoes from meteor showers. As meteors burn up in the atmosphere they produce ionized trails, which reflect radar signals. This discovery interested Bernard Lovell, of the University of Manchester, who was searching for similar echoes from the trails left by cosmic rays striking the atmosphere. As a pioneer in World War II aircraft radar development, Lovell had access to surplus radar equipment, which the University allowed him to park at their botany research station at Jodrell Bank, south of Manchester. (A peculiar coincidence: Jansky lived in a town called Red Bank, New Jersey; Lovell set up shop at Jodrell Bank, England; the U.S. National Radio Astronomy Observatory is located at Green Bank, West Virginia. This surfeit of banks in radio astronomy locations has still no reflection on the profession's remunerative benefits!)

Lovell's observations revealed no cosmic ray echoes, but more and more meteor trails, and for years meteor astronomy was a focus for research at Jodrell Bank. As the radio antennas grew in size, so did their potential for doing radio astronomy. Lovell subsequently propelled Britain into the forefront of the science by masterminding the construction of what was for a many years the world's largest fully steerable radio telescope, the 250-ft diameter Mark I, shown in a later incarnation in Figure 1.2. Completed just before the world's first artificial satellite, Sputnik I, was launched in 1957, the Mark I was the only radio telescope in the world capable of picking up radar echoes from the satellite's carrier rocket and played an important role in stimulating the United States to get more active in radio astronomy, and to develop a more effective radar system for national defense.

## 2.3. The Southern Skies

Observation of the southern skies fell to the Australian radio astronomers led by J. L. Pawsey, studying the sun, and J. G. Bolton, studying other radio sources, who also began by using surplus radar equipment. They invented a neat trick to make their radio antennas work more effectively. The resolution obtainable by a radio telescope (that is, its ability discern small-scale structure) depends on the diameter of the dish. If two or more antennas are used as an interferometer (see Chapter 1), the highest achievable resolution is determined by the maximum distance between them. Instead of building two antennas and spacing them hundreds of yards or miles apart, Bolton's team placed an antenna on top of a cliff by the ocean. For long wavelength radio waves the ocean's surface acts like a mirror. From the radio source's point of view, the cliff-top radio telescope consisted of two antennas, one on top of the cliff and the other apparently some distance below it, seen reflected off the water. These two antennas, one real and the other a reflection, acted together as an interferometer.

With this ingenious device, capable of 10 arcminutes of resolution, the Australians made some of the most important observations in early radio astronomy. They discovered that enhanced solar radio emission was associated with sunspots (in 1946) and that the temperature of the radio emitting regions of the sun were at a million degrees, a conclusion based on detecting radio waves from what turned out to be the solar atmosphere (called the corona) and studying the spectrum of these radiations (see Appendix A.4). They also observed the first solar radio burst (1947) produced by violent explosions known as flares. They confirmed the position of the Cygnus A radio source (1947) and found several new radio sources which helped to arouse the interest of optical astronomers in the new science. They also discovered the Taurus A radio source in 1947 (see Chapter 4 for its radiograph), and their accurate position measurements facilitated its identification with the Crab Nebula (1948). Then Centaurus A and Virgo A (Chapter 10) were added to the (at the time) very short list of radio sources associated with galaxies.

Observations of the radio emission from our own Milky Way galaxy (in 1952 and 1953) led to the discovery that there is a bright radio source in the constellation of Sagittarius (Chapter 5). At the time it was believed that the galactic center was about 32° away from this radio source, but subsequently, by international agreement in 1955, the location of Sgr A was taken to define the true galactic center.

## 2.4. Who Could Have Guessed?

In the beginning of 1950, the radio discoveries had barely made an impression on "normal" astronomers. The new breed of radio astronomers had discovered radio waves from the Milky Way and the Sun, and had managed to locate several radio sources that were optically identified. However, the picture appeared very confusing. Nearby galaxies, such as M31 in Andromeda, were at best faint radio sources while very distant galaxies, such as Virgo A (M87) and Cygnus A, were powerful emitters of radio signals. Centaurus A was associated with the galaxy NGC 5128 and was clearly not at all well behaved because it showed a dark dust lane crossing in front of an elliptical galaxy not expected to contain dust. At the time, no accepted theory existed to explain their radio emissions. In fact, around 1950

... radio astronomers were greatly impressed by the almost total lack of connection between radio observations and the visual sky. It did not seem impossible then that there were two separate kinds of celestial objects, each requiring distinct research techniques.

With regard to the problem of explaining the existence of the newly discovered radio sources, Greenstein commented:

No rational explanation that explains the weak (radio) emission from the brightest nearby galaxy, the Andromeda Nebula, can also apply to the faint distant radio source Cygnus A. You have to break down the prejudice that the world is pretty much as you know it, and begin to think of a world which is not like the world you understand.

Breaking through preconceived notions is something that has frustrated many a scientist (as well as philosopher, politician, or lay person). Who, at that time, could possibly have guessed at the amazing scenario that now accounts for the cosmic radio waves. Radio signals from the Milky Way are produced by cosmic ray electrons spiraling around magnetic fields stretched out in space between the stars. In the 1930s and 1940s no one knew that interstellar space contained cosmic ray electrons or that there were magnetic fields between the stars. At the time, cosmic rays were defined as protons (but not electrons) from space that struck the earth continuously. Cosmic ray physicists didn't concern themselves too much about the origin of the cosmic rays, nor did they know what happened to the electrons. Those researchers were mainly interested in studying the composition and physical properties of the particles that did reach their detectors. The absence of electrons was noted, but who would have thought that the electrons didn't reach the earth because they had wasted their energy radiating radio signals in interstellar space.

After the World War II, Enrico Fermi proposed that cosmic ray electrons could be accelerated in interstellar space, provided magnetic fields were present, but it wasn't until 1951 that evidence for such fields was obtained through the observation of the polarization of starlight by dust grains aligned by those fields. Later, when supernova remnants (the remains of exploded stars) were recognized to be strong sources of radio waves and when their polarization was measured, astronomers did realize that cosmic rays originated in supernovae. (Polarization refers to a preferred plane of vibration of the incoming radio waves, for example, horizontal or vertical or at some angle in between.) The cosmic ray electrons, spiraling about magnetic fields, cause the supernova remnants to shine. The electrons then leak out into space and ultimately cause entire Milky Way to glow with radio energy. Who, back in 1950, could ever have dreamt up something so outrageous?

## 2.5. Identity Crisis

During the 1950s more radio sources were discovered and cataloged, and arguments raged as to what the new data meant. The first generation of large radio dishes, the 250-ft diameter Jodrell Mark I (1957), and the NRAO 300-ft (1962) were years from completion. Receiver technology was still relatively crude and internally generated receiver noise hampered efforts to detect faint radio signals from space. (Receiver noise is generated because elements of the radio receiver are themselves at some finite, and in those days, quite high temperature.) By the late 1950s the science appeared to be in a state of relative confusion. David O. Edge and Michael J. Mulkay, who have traced the early development of radio astronomy in their book, Astronomy Transformed, observed that by 1958:

... we are .. at a time of maximum uncertainty and confusion in the history of work on radio sources. Agreement between the two major groups engaged in survey work (in Australia and Cambridge) is minimal, and the status of many of the observations is radically in doubt.

Argument raged and regarding the general state of radio astronomical knowledge at the time these authors ask:

... what was achieved, by 1958 ....? A handful of optical identifications, of an odd assortment of objects, 'normal' and 'abnormal'; a suggested mechanism for radio emission from some of these (this being very largely the work of optical astronomers and theorists); a growing realization (many having already realized it quite early in the fifties) that the majority of radio sources must be extragalactic .....; catalogs of sources numbering, for all the hopes, merely hundreds, and those still the subject of controversy; some (but not many) radio diameters, spectra, and a few polarization measures; cosmological claims radically in doubt; source counts in complete disarray...

In this quote, the term "source counts" refers to the number of radio sources observed at ever fainter and fainter levels. At the time these counts were believed to be a potential holy grail that would allow us to understand whether the universe began with a bang or existed forever (see Chapter 13), an expectation that was never realized.

Radio astronomers, who were not considered by traditional stargazers to be astronomers until the late 1950s and early 1960s, had clearly stumbled into a new universe, an invisible universe. Much like a blind man has to learn his way around his world, the radio astronomer not only had to develop new ways of sensing what was out there, but also had to invent methods for communicating what it was they were discovering. After careful discussion with optical astronomers, the radio astronomers attempted to infer whether their observations related to something known to other astronomers, or whether they were sensing a completely different universe.

The authors of *Astronomy Transformed* suggested that radio astronomy went through several stages. The first stage began with the discovery of radio waves from objects like the sun and the early exploration of these discoveries. There was a sharing of information between several groups and by the end of this stage (early 1950s) it was recognized that there were several astronomical lines of inquiry involved:

During the ensuing stage radio astronomers publish increasingly in optical journals, join optical (astronomy) societies . . . . and come to hold joint conferences with optical astronomers. Essentially a bond is formed with the 'real' astronomers. The radio technology is developed so that good data, which make sense and are repeatable, are generated.

## 2.6. An Epoch of Discovery

It wasn't until the 1960s that the bond with "real" astronomers began to be forged on a large scale, following stunning new discoveries made possible by enormous improvements in receiver technology and the construction of large reflector-type radio dishes and interferometers with ever greater baselines, the distance between their individual dishes. These contributed to the radio astronomer's ability to measure radio source positions with greater accuracy, sufficient to force the attention of the general community of optical astronomers. The days when an old-timer at a meeting of the Royal Astronomical Society in London was overheard to ask "What is this new-fangled wireless astronomy?" were past.

The 1960s also saw the transformation of radio astronomy into a "big science," which brought with it a remarkable period of exciting new discovery. Research at the forefront was, however,

. . . only open to those groups with sufficient expertise to develop the complex techniques required and with sufficient repute to attract extensive financial support from government and from industry.

Radio astronomy growth during this phase largely bypassed the United States. It was only in the mid-1960s, following the Sputnik panic that urged greater emphasis on science, that US radio astronomers began to catch up.

The 1960s and early 1970s saw the discovery of quasars, pulsars, radio source polarization, complex interstellar molecules, interstellar masers, radio stars, bipolar flows, radio jets, and extragalactic molecules, and the first measurement of the interstellar magnetic field strength. Those years also saw the solidifying of the theoretical understanding of the emission mechanisms involved in thermal and nonthermal radio sources (Appendix A.4), while explanations for the maser mechanism as well as pulsar radiation (Chapters 7 and 8) were quick to develop.

According to Edge and Mulkay;

Stage three is characterized by a growing concern with astrophysical problems, arising largely from the major discoveries of quasars and pulsars and from the advent of new approaches like those of ultra-violet, X-ray, and infra-red astronomy. By this stage radio methods have become an established part of astronomy.

By the mid-1960s and certainly at the end of that decade, it was firmly demonstrated that the universe was not as quiet as had long been assumed. The universe is wracked with violence on all scales ranging from exploding stars to exploding galaxies and quasars and even to violence on the scale of the universe itself, the Big Bang.

From the point of view of growth, and availability of funds to drive this growth, the period 1960–1975 might be called the Golden Age of radio astronomy. That was also when radio astronomical terms such as quasar and pulsar entered the mainstream vocabulary.

---

Whenever someone hears that I am a radio astronomer, and after they have passed through the phase of confusing this with some form of astrology, I am often asked "How far can the telescope see things?" Bearing in mind that radio telescopes are not something you can see through, we nevertheless use the colloquialism of "seeing" radio waves. That is part of the jargon of the trade. We can't say we listen to radio signals from space either, because there is nothing to hear that the human ear can detect against the background "noise" produced by the radio receivers attached to the radio telescope. (It is a sign of the times that a radio telescope is best described as a large satellite dish!) Instead, we look at the output of a computer program that converts the radio signals generated by a host of interesting physical events in the depths of space into numbers or maps of what those objects would look like if you could literally "see" radio signals.

As regards the question "How far can a radio telescope see?" I usually respond that they can "see" farther than the Hubble Space telescope, very nearly to the beginning of the universe. The reason is simple. Virtually every radio telescope ever constructed, if equipped with a suitably sensitive receiver, could, in principle, detect the faintest of whispers left over from the Big Bang (Chapter 13), provided reception is not swamped by terrestrial signals (interference) that might overwhelm anything that reaches those dishes from outer space.

---

## References

1. K. Kellermann and B. Sheets (eds.), *Serendipitous Discovery in Radio Astronomy*, National Radio Astronomy Observatory, Green Bank, WV, 1983.
2. D. O. Edge and M. J. Mulkay, *Astronomy Transformed*, Wiley-Interscience, London, 1976.

# 3
# The Radio Sun and Planets

## 3.1. War Secrets

On the morning of February 12, 1942, the German battle cruisers Scharnhorst and Gneisenau passed undetected through the English Channel on a voyage from Brest in France to Kiel in Germany. The reason they sailed unmolested by British warplanes was that the British radar was being jammed by radio interference. J. S. Hey, a physicist who had learned something about newly invented radar at the outbreak of World War II, was assigned the task of investigating the jamming. The suspicion was that the Germans had come up with a device that could blind the British radar.

A few weeks later widespread jamming again occurred and the military responded by going on extreme alert, yet no hostile action followed. Hey discovered that jamming had occurred only in the daytime, especially when the sun rose in the east and the radar antennas were pointing in its general direction. A check with the Royal Observatory at Greenwich revealed that at the same time a large sunspot group was visible on the solar surface. Like Jansky, before him, Hey found that extraterrestrial radio signals were responsible for unwanted "interference", and so he wrote a secret memo reporting that the jamming seemed to be produced by radio signals from the sun. Thus, out of the desperate situation of World War II, the seeds of solar radio astronomy were sown.

## 3.2. The Plasma Sun

The sun is a churning mass of hot ionized gas with magnetic fields threading their way through every pore and core, driven by energies boiling out from the interior where the fusion of hydrogen into helium at a temperature of 15 million K liberates the nuclear energy that keeps the cauldron boiling. [The symbol K denotes the temperature in degree Kelvin, which begins with 0 K at absolute zero ($-273°$C). On this scale the temperature of ice ($0°$C) is 273 K.] Heat radiates toward the solar surface, which maintains a temperature of about 6,000 K. A gas at this temperature radiates primarily light (and some heat and ultraviolet radiation,

23

as we know from personal experience while sunbathing) by the thermal emission process.

A cloud of hot ionized gas is known as plasma, and it can support a large number of wave motions within its volume. When ionized particles move in harmony they begin to radiate energy, and the generation of radio waves by the solar plasma oscillations is of fundamental importance to the understanding of solar radio waves. The level of this understanding has reached such awesome proportions that one prominent solar physicist has suggested that the theory of solar radio emission is now so well developed that most astronomers can no longer understand it! This need not concern us, because we will only touch upon the most important phenomena and leave the details to the experts.

The visible surface of the sun, known as the photosphere, is mottled by light and dark patterns where plasma actively surges up and down, cooling as it rises, heating as it falls. Occasionally small regions become unusually active. Magnetic fields tangle and knot and that can trigger more upheavals. The fields tear, twist, and turn, and bits and pieces intermingle and reconnect to form new patterns of force. The reconnection of magnetic fields is usually accompanied by the sudden release of vast amounts of energy—energy that was originally held in the fields and is then converted into the explosive ejection of particles into space. These explosions are observed as bright flares of light on the solar surface, often near cool sunspots, where magnetic field activity is particularly intense. The plasma around these magnetic field explosions is set into oscillation and radio waves are generated that travel outward to reach the earth 8 min later. (The sun–earth distance is 8 light-min.) When sunspot activity is very great, and solar flares repeatedly burst out over the surface, intense radio noise is produced.

During active spells, solar magnetic fields coil and uncoil, heave and churn, and arch upwards. These arches are called prominences when seen at the edge of the sun. They rear up like uncoiling snakes, and great clouds of plasma arc offered a way to escape the boiling heat below. Tentacles of magnetic field break and release their grip and clouds of particles stream out into space, triggering oscillations in the surrounding plasma as they rise into the solar corona.

The moving clouds successively trigger radio emission higher and higher in the corona, and occasionally clouds of ejected plasma may reach the earth and impinge on its magnetic field. The earth's field acts as a shield, an invisible force field, which protects us from the solar particle storms. The traveling plasma clouds thus slide past the planet, leaving spaceship earth untouched by their harmful effects, which would result if the plasma clouds were to crash unimpeded into our atmosphere. These high-energy particles can destroy ozone (the molecule $O_3$) that exists high in the atmosphere and protects us from direct solar ultraviolet radiation. Ozone absorbs ultraviolet radiation, which is fortunate for us, because large doses are fatal to terrestrial life forms.

After particularly violent solar storms, particles can penetrate the protective terrestrial magnetic field, especially in the region of the earth's magnetic tail, which stretches out beyond our planet like the wake behind a boat. The tail is swept there by the perpetual wind of particles blowing out of the sun. Following a solar storm, these particles (mostly electrons) may worm their way into the earth's magnetic tail, where they promptly rush helter-skelter along the magnetic field toward the polar regions of our planet. These electron streams then crash violently into the highest regions of the terrestrial atmosphere where they collide with, and ionize, atoms of oxygen and nitrogen. These gases then vibrate with energy so that they produce the magnificent fiery displays known as aurorae.

## 3.3.  Solar Radio Emission

The study of solar radio waves was launched in earnest in the postwar years, when many physicists the world over salvaged surplus radar equipment whose antennas and receivers were ideal for studying the sun. Today the radio sun has been observed across the radio spectrum and modem solar radio astronomy is replete with extraordinary detail and we can mention only some of the highlights.

The sun emits radio signals through the synchrotron process, which involves high-speed electrons spiraling around magnetic fields, as well as thermal radiation from the hot plasma, which is produced by virtue of the motion of the electrons in the plasma. A third mechanism, which has many variations, involves natural oscillations of the plasma itself. Radio emission can occur at the frequency of the plasma oscillations as well as multiples of this frequency.

## 3.4.  The Quiet Sun

Radio emission from the quiet sun is observed at times of sunspot minimum and comes from regions low in the corona. A slowly varying component may be observed which varies with the ponderous rotation of the sun, one cycle after every 28 days. The variable intensity is partly related to the presence of cooler regions known as coronal holes, which alternate with a slightly warmer, more normal plasma over the solar surface. The quiet sun, by definition, is observed when there is little violent activity occurring.

The slowly varying component is also related to the presence of filaments of hotter gas, which thread their way over the solar disk. In their immediate neighborhood, temperatures change from 6,000 K in the filaments to 2 million K in the surrounding corona, which has long been known to be extraordinarily hot for reasons that remain mysterious.

## 3.5.  Solar Radio Bursts

On a regular basis, especially when many sunspots are present, regions on the surface of the sun may grow steadily hotter and brighter until a flare explosion occurs. Near such active regions, whether or not accompanied by a flare activity, bursts of radio waves may be generated whose variations in frequency and overtime can be very complex.

## 3.6. Radio Signals from the Planets

All objects at "everyday" temperatures emit radio waves. This includes the moon, earth, planets, and your own body. The dark universe beyond the stars is at a temperature of 2.7 K (see Chapter 13). The earth, at a temperature of about 290 K, would appear as a thermal radio source to a distant radio astronomer on Pluto. The fact that all objects at a finite temperature emit radio waves by the thermal process means that even if a radio telescope is pointed at the ground or at a distant clump of trees, it will pick up radio waves.

Radio astronomers expected thermal emission from the planets, and quickly Mercury, Mars, and our moon were found to be relatively normal sources of radio emission, their radio brightness depending on frequency as expected from objects at their particular temperature. (The temperatures of the planets are directly inferred from heat, or infrared, measurements.) Venus, however, produced a surprise. Its cloud tops are at a temperature of 230 K, but in 1956 the first radio observations of this planet showed that its temperature was 600 K. This discovery came as a considerable shock. It turns out that the Venus cloud layers, filled with carbon dioxide, act as a greenhouse, keeping the surface of the planet at 600 K, a temperature later confirmed by direct measurements from landing spacecraft.

## 3.7. Jupiter's Radio Bursts

In 1955, as part of a 22-MHz sky survey, two budding radio astronomers (B. F. Burke and K. L. Franklin, working at the Carnegie Institute in Washington, D.C.) made daily observations of the Crab nebula supernova remnant, which is a strong radio source named Taurus A. They used it to produce a standard signal to calibrate their data. As they further developed their antenna and receiver, they persisted in monitoring the Crab. When the time came to begin the systematic search for new radio sources, they had to make daily adjustments to the antenna system so that it would receive signals from directions a little further south each day.

The arbitrary decision to start their mapping program by pointing the telescope further south (rather than north) paved the way for their major discovery. Unknown to them, Jupiter was also lurking up there and it was moving a little further south with respect to the stars each day. Soon their data began to reveal unwanted "interference." This interference came through soon after the Crab nebula was observed. A few days later, they began to take the signals seriously and sought an explanation. A colleague, Howard Tatel, apparently jokingly, suggested it was Jupiter.

On that same evening, out in the field where the antennas were located, Burke noticed a bright object in the twilight sky and asked his partner what it might be.

"Jupiter," came Franklin's answer, causing them to laugh at the odd coincidence in view of the remark by Tatel earlier in the day. Neither of them noticed that Jupiter was in Gemini, the constellation immediately adjacent to Taurus, the home of the Crab nebula.

The next day Franklin, perhaps in desperation, decided to explore the Jupiter connection more closely. To his complete surprise, he found that, indeed, Jupiter could be blamed for the "interference." Jupiter's radio signals turned out to be not steady emissions, such as might be produced by the thermal process, but intense bursts, not unlike those produced by the sun. This was one of the most unexpected discoveries in radio astronomy. By chance, the peak energy in the radio bursts is concentrated in the radio band around 20 MHz. If Burke and Franklin had been observing at 40 MHz or higher, or at another time of year, or if they had started to survey to the north of the Crab nebula, the radio bursts would not have been discovered for years.

The story had an ironic twist. Australian radio astronomers had been observing the radio sky at 19 MHz and years before had noticed a peculiar source of radio emission, but its cause had remained a mystery to them. Their antennas did not have sufficient resolution to pinpoint the source, and privately they believed that perhaps the swishing sounds they heard were terrestrial interference originating somewhere over Indonesia. With the announcement of Burke and Franklin's discovery, the Australian researchers looked back at their old records and found that Jupiter had also produced their "interference" and that its signals were visible on records going back 5 years. Thus, within weeks of the discovery that Jupiter was a radio source, they had 5 years of data with which to work.

Even with a sensitive, easy to build, and cheap radio receiver and antenna, these bursts can be heard on a loudspeaker.

In 1964 it was discovered that one of Jupiter's large satellites, its moon Io, plays a role in determining when the burst radiation is beamed in our direction. The emission mechanism is complicated, barely understood, and apparently related to plasma wave phenomena, which in turn are related to the location of Io with respect to the tilted magnetic field of Jupiter. The radio-burst radiation is beamed along narrow angles, so that when the beams sweep by the earth, the signals are observed to vary in intensity. Whether or not the burst radiation is received on earth depends on where the earth is in Jupiter's sky and on Io's location with respect to the Jupiter–earth line.

Then Jupiter provided radio astronomers with another considerable shock.

## 3.8. Jupiter's Radiation Belts

Jupiter, the largest planet in the solar system, is very cold because it is very far from the sun. Infrared studies of its cloud tops had measured a temperature of 150 K (as compared to 220 K for the earth's cloud tops). Radio observations at 10,000 MHz confirmed that Jupiter behaved like a thermal radio source at this temperature. Despite the discovery of the radio bursts, which were clearly being generated by some nonthermal process, Jupiter was expected to be a normal thermal emitter at other wavelengths as well. However, at 3,000 MHz Jupiter's radio brightness implied a temperature of 600 K, and at 400 MHz it was 70,000 K. Clearly, Jupiter was far more than a simple thermal source. Furthermore, these radio signals were found to be polarized, a sure sign that it was also a nonthermal radio source, behaving somewhat like distant radio sources in which cosmic-ray electrons spiral around magnetic fields to produce radio waves. No one had expected this.

Figure 3.1 shows a radiograph of Jupiter, which confirms the theory that the radiation belts, which reach from about 90,000 to 200,000 km above its cloud surface, are the source of the radio waves. The radio waves are polarized, and because Jupiter's magnetic field is tilted by 10° with respect to its rotation axis, the intensity of the received radio signals varies with time and as the direction

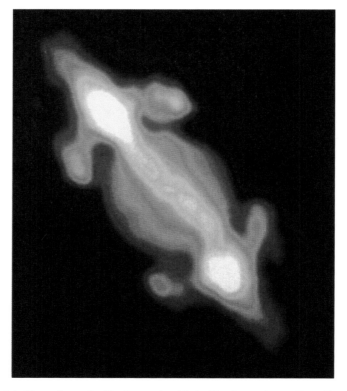

FIGURE 3.1. The radiograph of Jupiter made with the Very Large Array. The radio emission comes from its atmosphere emitting thermal radiation related to its 150 K temperature and from the nonthermal emission from the radiation belts that surround the planet. Investigator: I. de Pater. (Image courtesy of NRAO/AUI.)

of polarization of the incoming radio waves rocks back and forth it is possible to accurately determine the planet's rotation period. The Jovian day is 9 h 55 min and 29.71 s long.

## 3.9.  The Planets as Radio Sources

Saturn and the earth also generate radio emission similar to the Jupiter bursts. The peak of Saturn's burst radiation occurs at around 500 kHz and the earth's at around 60 Hz.

To a radio astronomer at the outer reaches of the solar system, the earth would appear as the strongest radio source in the sky at 60 Hz. Those radio signals originate in the auroral regions, but the signals are not observable down here on the earth's surface because such radio waves cannot penetrate the electrically charged ionosphere in the upper reaches of our atmosphere. It is an odd coincidence that worldwide sources of commercial alternating current are distributed at 50 or 60 Hz, a frequency which happens to be at the natural frequency of the earth's burst radiation.

## 3.10.  Planetary Radar

A variation in the usual radio astronomy way of doing business is to transmit a radio signal at a distant object and listen for an echo. Known a radar astronomy, one of the world's most powerful radar transmitters is sometimes used with the Arecibo radio telescope and the echoes from the moon, asteroids, or a planet such as Venus are gathered and processed to produce reconstructed images of the surface of those objects. Figure 3.2 shows a radar map of the Orientale basin on the western edge of the moon's earth-facing side. This image was obtained by transmitting a radio signal from the Arecibo radio telescope and receiving the faint echoes with the 100-meter Robert Byrd telescope in Green Bank.

---

As a teenager I used to fantasize about being able to walk on the moon some day so that I would know what was up there. In early 1966 I experienced something that was a close second to that wish.

On February 3, 1966, a Soviet spacecraft (Lunik 9, a.k.a. Luna 9) made the first ever-successful soft landing on the moon. At that time I was on the staff at Jodrell Bank and its director, Sir Bernard Lovell, had over many years developed close ties with Soviet space scientists. The giant radio telescope was often used to track their spacecraft and to record their signals as a backup for the Soviets, should their ground stations have a failure at a critical moment in their mission. Lovell's contacts in Russia informed him in advance of the frequencies they planned to use for any given space mission.

The Lunik 9 mission was no exception. Lovell knew of its intended landing and the frequencies of its transmissions, and he was in his lab taping the signals as the spacecraft landed on the moon. The Russians had again beaten the United States to a space first. The scene was now set for what turned out to be one of the most dramatic events in the space program and certainly in my life.

As the giant 250-ft dish slowly tracked the moon across the sky on that historic day, the halls of the observatory were crowded with reporters from all over the world. Most of the staff, myself included, viewed the whole event with a jaundiced eye. What did the reporters think they would learn? Late in the afternoon I negotiated my way through the crowds of reporters and went home. When I returned the next morning the reporters were gone and I got to hear about some amazing action during the night. One of the senior staff members, J. G. Davies, had been listening to the signals from the moon's surface and recognized the sound of a fax machine! In those days fax machines were exclusively found in newspaper offices and the reason that Davies had heard this before was that several years previously the Russians and Jodrell Bank had performed a joint experiment in which fax pictures had been bounced off the moon between England and a radio telescope in the Crimea using radar to carry the signals.

As a long shot, Lovell managed to convince The Daily Express, a London newspaper, to load up a fax machine, about the size of a small refrigerator, and immediately truck it up to Jodrell. In the early hours of the morning the tape of the recoded signals had been played into the fax machine but had produced no images.

Around noon that day we (and I include dozens of reporters who had returned to the scene) watched a European wide television program from Moscow. It was expected that we would see pictures of the surface of the moon. However, other than speeches by dignitaries and cosmonauts, no pictures were forthcoming. Also, no excuses were made. Clearly the Russians had nothing to show for their lunar landing.

Sir Bernard Lovell then told the assembled multitude that when the moon rose again that day, in an hour or so, they would feed any signals from Lunik 9 directly into the fax machine in case that worked. Nothing would be lost by trying.

At this point I stuck my nose into the area where the space tracking receivers were located and where the Daily Express technicians had set up their fax machine and a portable darkroom—a small tent.

The moon rose and the telescope began to track its motion in the eastern sky. We heard the signal and the drum on the fax machine that carried the photographic sheet began to rotate in response to electrical energy that had traveled 235,000 miles to get there.

The first picture took about two and a half minutes to come through and was quickly processed by the technician, but the image was half white and half black, a total disappointment.

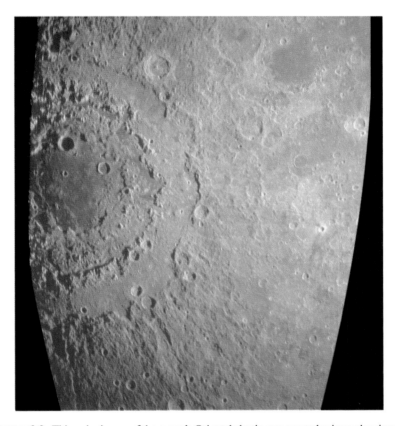

FIGURE 3.2. This radar image of the moon's Orientale basin was created using radar signals transmitted from the Arecibo Observatory in Puerto Rico while the echoes were received using the Green Bank Telescope of the National Radio Astronomy Observatory in West Virginia. The radar signal had a wavelength of 430 MHz and could penetrate to depths of several meters in the lunar surface. Bright areas are due to slopes that face toward the radar signal (such as crater walls or mountains) and to rocks at or just below the surface. Dark areas are typically associated with smooth lava flows that fill large craters and basins on the Moon. Investigator: Bruce A. Campbell. (Image courtesy of NRAO/AUI and Bruce Campbell/Smithsonian Institution.)

Clearly the Lunik 9 mission had suffered a major setback and so Lovell informed everyone that he would give a closing statement in the conference room and all the reporters filed out.

I, together with two technicians from the Daily Express—one for the fax machine and the other to develop the pictures—and a Jodrell Bank technician in charge of the receiving equipment, stayed back. Then the fax machine again began to respond to a signal from the moon. Again the photographic film was moved to the darkroom. I had taken up a position by the tent and I heard the man inside mutter, "Mmm, we seem to have something here."

He held open the tent flap and let me peer inside and there drifting in the tray of fixer was the world's first close-up photograph the rock-strewn surface of the moon! My immortal words on seeing this vision were never recorded!

Word then reached Lovell who instructed all the reporters to wait in the nearby lobby while he went to see what had happened. He looked at the photograph and made an equally immortal comment that went equally unrecorded, as I was later fortunate enough to lament to Neil Armstrong on the 1973 Eclipse Cruise on board the Canberra where we both happened to be lecturing.

Other pictures followed with equal clarity and the question arose as to what should be done with the photos. The essence of the problem was that these were Russian space program pictures and should a British newspaper, with the aid of a British observatory, be the first one to publish them? If so, which newspaper had the rights, if any, to the photos? Were these photos in the public domain or were they the sole copyright of the Daily Express?

It would have been impossible to stop the Daily Express from publishing the photograph, but when they insisted on exclusive rights to all the images (which were part of a panoramic scan of the lunar horizon), there was nearly a riot among the reporters. Sir Bernard Lovell came up with a brilliant British compromise. The first picture received was distributed to everyone. The rest were to be exclusive to the Daily Express which the next morning rightly proclaimed the "Scoop of the Century." He had no choice. He was subsequently criticized by scientists in the United States (none of whom had equipment to make their own fax images!) who said that he should not have released the pictures before the Soviets did. But all of us at Jodrell Bank figured that something had gone wrong in Russia, which is why nothing was shown on their Eurovison TV program.

To this day I do not know whether the Russian space tracking facility had their pictures developed as fast as ours. The first image received at Jodrell was the first photograph obtained by the Russians. I may have been the first person (other than the darkroom technician working in his safe light enclosure) to see a photo from the surface of any other object in the solar system.

My childhood desire to go to the moon will never be fulfilled, but at least this moment was as close as I could have imagined in my wildest dreams. The moral of the story is that my curiosity is what drove me to be standing there while the photo was being processed.

We discovered the next day that the pictures received at Jodrell Bank were elongated in one direction due to a gear change the Russian engineers had made in their fax machine. Yes, the fax machine that they landed on the moon. To this day I am amazed that Lunik 9 carried a piece of hardware that was widely available in the West. They basically threw it very hard from the earth and landed it cleverly and safely on the moon.

# 4
# The Galactic Radio Nebulae

## 4.1. The Supernova—Stardeath

On a chi-chou day in the fifth month of the first year of the Chih-ho reign period a guest star appeared in the south east of Thien-Kuan measuring several inches. After more than a year it faded.[1]

With these words Chinese astronomers in the year 1054 AD recorded their observation of a new star in the constellation Taurus. Early in the morning of that same day, July 5, Plains Indians in the Western United States witnessed the appearance of this new star—an event so startling to them that they etched a record of it into the rocks. A pictograph found in Chaco Canyon, New Mexico, shows a star-like symbol, seldom used by those Indians, drawn next to a crescent (moon). The new star described by the Chinese astronomers, who had a long tradition of observing the heavens behind them, did indeed appear close to the crescent moon and it would have been visible in the early morning from the rocky overhang in Chaco Canyon.

Today we call such a guest star a supernova, and recognize the phenomenon as the violent destruction of a star in its final minutes. The remains of the dying star of 1054, which was visible during the day for several months, are now viewed as the supernova remnant known to optical astronomers as the Crab Nebula and to radio astronomers as Taurus A, one of the brightest radio sources in the sky. The radio portrait of Taurus A shown in Figure 4.1 reveals a display of luminous filaments that emit not just radio energy but also X rays, ultraviolet, and light. (Note that the appellation "nebula" refers to any cloud-like object in space. All nebulae look like fuzzy clouds when viewed through small telescopes.)

The Taurus A radio source had already been associated with the Crab nebula in 1947 because it was quite obvious in photographs and it became a primary testing ground for the theory of synchrotron emission. The light and radio waves from the Crab are polarized, and its spectrum is clearly nonthermal. Today, more than 950 years after the explosion, the Crab continues to radiate because of energy injected

---

[1] Ho Peng Yoke, *Vistas in Astronomy*, Pergamon Press, 1962, page 127.

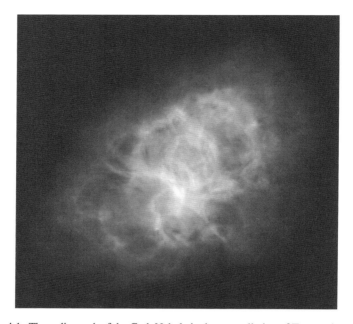

FIGURE 4.1. The radiograph of the Crab Nebula in the constellation of Taurus, the remnant of the supernova in 1054 AD that was observed as a "guest star" by ancient Chinese astronomers as well as Plains Indians in North America. Located about 6,000 light-years from earth, its filamentary features are expanding at about 1,000 km/s, driven by both the energy of the original explosion and a powerful wind from a pulsar (Chapter 8) at its center. Investigator: M. Bietenholz. (Very Large Array image courtesy of NRAO/AUI and M. Bietenholz.)

into the nebula by a very odd star that spins furiously at its center. This object, known as a pulsar, will be described in Chapter 8.

## 4.2. Recent "Guest Stars"

In late August of 1975 a new star, almost as bright as Polaris, the Pole Star, appeared just beyond the tail of Cygnus the Swan. Eight days later it faded from sight, and by then it was officially called Nova Cygni 1975. It wasn't a supernova, but something less dramatic—a nova, which means "new star." Unlike a supernova, the complete disintegration of a star at the end of its life, a nova is a relatively gentle explosion, which tosses a cloud of gas away from the star's surface. Some novae repeatedly convulse and eject hot clouds of gas that cause the star to appear a million times brighter for a week or so. Then the nova fades to its normal brightness over the next few decades. Novae are, at best, weak radio sources as a result of thermal emission from the hot gas of the star's outer layers.

A supernova, however, shines with the light of a billion suns and over several months slowly fades away, although its remnant may still be visible to large telescopes for thousands of years afterwards.

Not since the invention of the telescope (in 1609) has a supernova explosion occurred in our Galaxy, at least one that we are aware of. A distant star could have blown up thousands of years ago if it were located half way to the galactic center and its light wouldn't have reached us yet! In 1604 Johannes Kepler, the astronomer famous for discovering the laws that govern the movement of the planets about the sun, did record the appearance of a supernova visible to the naked eye.

One supernova per 50 to 100 years in a typical Galaxy is estimated to be a reasonable average rate for such events, based on observations of hundreds of supernovae in other galaxies. The Milky Way is long overdue for its next one. In 1987 a supernova was spotted in the Large Magellanic Cloud. Lucky southern hemisphere observers could watch as it faded over months and its remnant continues to be closely monitored by astronomers at wavelengths across the electromagnetic spectrum from radio to X rays.

## 4.3. Cassiopeia A

The brightest radio source in our heavens, other than the sun (which appears bright because it is so close to us), is a supernova remnant in the constellation Cassiopeia. First efforts to identify an optical object at its location were rather disappointing. Nothing as blazingly obvious as the Crab Nebula was found. Instead, very faint filamentary material is seen on photographs, the only visible remains of an exploded star. The presence of dust between the sun and this remnant, which is located 10,000 light-years away, may be preventing us from seeing the object. Radio waves ignore such dust and have revealed Cas A to be a glorious display of radio emitting filaments. The radiograph of Cas A shown in Figure 4.2 is one of the most stunning radiographs ever made. The image appears ring-like, suggesting a shell of material ejected by an explosion. Observations of motion in the filaments indicate that the explosion must have occurred in 1680 AD, but no record exists of anyone having seen it back then.

## 4.4. Supernovae of Type I and Type II

The theory of supernovae is reasonably well developed. Fortunately, the sun is not likely to suddenly explode violently to wipe out all terrestrial life. On the other hand, it is certain that when the sun enters its dotage, somewhere between 2 to 5 billion years from now (depending whose estimate one adopts—neither of which should give us cause for concern), the earth's atmosphere will be destroyed.

Stars containing more than four solar masses of gas are likely explode at the end of their lives, but not all of them will do so in the same way. The Tau A and Cas A represent a different class of event compared to Kepler's supernovae, or the one seen with the naked eye in 1572 by another famous astronomer of old,

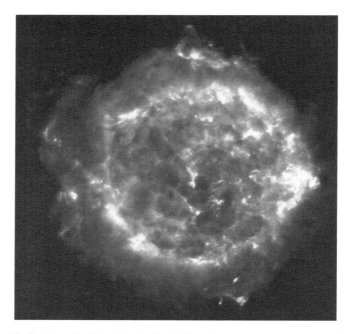

FIGURE 4.2. Radiograph of the Cassiopeia A (Cas A) supernova remnant created by a star that exploded in 1680, the date inferred from the current rate of expansion of the nebula. Investigators: P.E. Angerhofer, R. Braun, S.F. Gull, R.A. Perley, and R.J. Tuffs. (VLA image courtesy of NRAO/AUI.)

Tycho Brahe. The latter two are examples of Type I supernovae, believed to be the destruction of what was originally a white dwarf star—a highly mature star which, in its old age, shrinks to a mere shadow of its former self. These white dwarfs were probably members of binary star systems. (Approximately half the stars in our Galaxy are paired in binaries, unlike our sun, which has no close companion.) The interaction between binary stars can be very dramatic. If one member is a dwarf star, evolving slowly, and the other a more massive star, aging rapidly, the larger star may enter a phase in its life when it swells to enormous size. Some of its material then falls onto the dwarf and if this process continues for long enough the white dwarf may suddenly become incapable of absorbing any more of its neighbor's debris. Its surface layers overheat and explode outward. The Tycho and Kepler supernovae are believed to have been produced in this manner. Study of their light gives much information about the chemical constituents of the exploding material. Type I supernovae are regularly observed to occur in distant elliptical galaxies, which contain no interstellar matter, nor any young, massive stars are capable of becoming supernovae on their own. A Type II supernova, on the other hand, may involve the explosion of a single massive star due to a catastrophic increase in the amount of heat generated in its core as part of its "normal" evolution.

## 4.5.  Supernovae and Life

The filaments within supernova remnants produce nonthermal radiation. As the remnant expands, ages, and runs out of energy, the electrons responsible for that emission escape into surrounding space, where they become cosmic rays. These cosmic rays fill the Galaxy, all the while encountering magnetic fields between the stars. The electrons gain energy from such encounters and then, rejuvenated, transmit nonthermal radiation that we pick up as radio waves from the Milky Way (Chapter 5). The heavier particles, such as the protons created in the supernova explosions, may later strike earth as cosmic rays.

The total energy generated by supernova explosions over billions of years provides enough energy to propel interstellar clouds hither and thither, causing them to collide with each other. Supernovae keep interstellar matter stirred up, and when clouds collide the birth of a new star is triggered. It is likely that a nearby supernova remnant probably spawned the formation of the solar system.

Supernovae play a direct role in assuring our existence. The early universe contained none of the heavy elements, the basic constituents of matter, which make up our world. However, tantalizingly, the Hubble Space Telescope is revealing the products of stellar death in the very earliest galaxies seen at the edge of space–time. Those elements include oxygen, nitrogen, and carbon, elements essential for life, as we know it. All the atoms of which we are made (except hydrogen) were formed in stars and then injected into space and made available for future planetary formation because stars exploded. The heaviest elements, such as the gold and silver in rings or pendants, were formed during the supernova explosions themselves.

Today the material within the supernova remnant Cas A is being fed into space and sometime, somewhere, in the very distant future, some of it now radiating radio and light signals at us may be incorporated into alien living entities.

---

In imagination I often wonder what it would be like to live on another planet that happens to be located within 30 light-years of a star that explodes as a supernova. Imagine a technologically sophisticated civilization like ours. One day a new star appears that signifies the initial flash of the explosion. Within days a fatal dose of gamma rays and X rays from the event causes widespread death and triggers mutations that lead to radiation sickness. Knowing the exploded star is 30 light-years away and knowing how fast the shell of ejected matter is streaming out into space, their scientists estimate that they have about 300 years before the shock waves carrying lethal doses of high-energy particles strike their planetary system. There is only one thing to do and that is to plan to live underground where they have some protection from the cosmic rays and the X rays generated in the nebula. And that is where they will have to stay for thousands of years while the nebula envelops them. It seems likely that the sun formed as the result of the compression of interstellar matter triggered by a nearby supernova some 5 billion years ago. Fortunately there is no likely candidate star that might become a supernova within a thousand light-years of earth, assuming astronomers know enough about stellar evolution for us to be comforted by this conclusion!

## 4.6. Emission Nebulae—Star Birth

In the constellation of Orion, visible in the mid-winter sky in the northern hemisphere, the famous hunter of the heavens presents three equally bright stars lined up to indicate his belt. Just below it you may see three points of light outlining the sword hanging from Orion's belt. When viewed through binoculars the central star is revealed to be a faint, fuzzy nebula. Located 1,500 light-years away, this nebula glows because it has been heated to incandescence by ultraviolet radiation from four young stars called the Trapezium. These stars were spawned from an interstellar gas-and-dust cloud about a million or so years ago, when primitive Homo sapiens roamed our planet. The radio image of the Orion nebula, known as Orion A to radio astronomers, is shown in Figure 4.3. It is the prototype of a class

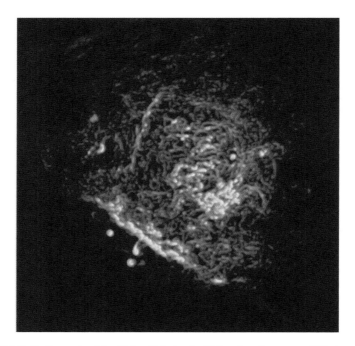

FIGURE 4.3. Radiograph of the Orion Nebula, the Orion A radio source. This image was produced by filtering out the large-scale structure to show only a tangled web of filaments. Investigator: F. Yusef-Zadeh. (Very Large Array image courtesy of NRAO/AUI.)

of radio source called an emission nebula. The gas in the nebula has been heated to incandescence by ultraviolet radiation from young stars within it and this hot gas, at a temperature of 8,000 K, radiates light, heat, and radio waves by the thermal emission process.

## 4.7.  HII Regions

Radio astronomers generally refer to emission nebulae as HII regions, where the symbol "HII" refers to ionized hydrogen. The symbol HI is used to represent cold or neutral hydrogen atoms, each of which consists of a single electron orbiting a proton. Hydrogen clouds exist everywhere in interstellar space (Chapter 6). It is within such clouds that stars are spawned. When hot stars begin to shine, their highly energetic ultraviolet radiation streams outward and when a UV photon strikes a hydrogen atom it can kick the electron hard enough to escape the powerful grip of the proton. The hydrogen atom is ionized and its constituent proton and electron now wander about freely. A large cloud made of a mix of protons and electrons forms an HII region that emits thermal radiation.

In the interstellar gas clouds in which stars are born, the larger, hotter stars literally eat their way through their surrounding gas cocoons, converting what had been cold matter into hot gases, which radiate their own energy. Within large emission nebulae many small, compact HII regions can be formed, each associated with a newly born star. The difference between a supernova remnant and an HII region is revealed by measurements of their spectra and polarization. The HII region has a thermal spectrum and its radio emission is unpolarized, while the radiation from supernovae is nonthermal and polarized.

## 4.8.  Planetary Nebulae

Other galactic nebulae that are radio sources include planetary nebulae, which represent far less violent affairs than supernovae. The supernova is the death of a massive star, while the planetary nebula signifies the death of a smaller, more normal star. The central star of the planetary nebula merely shrugs off its outer layers as it ages. What is left of the inner parts of the star may collapse to become a white dwarf and in due course will fade from view.

The existence of emission nebulae, supernovae, and planetary nebulae are constant reminders that we live in an extraordinarily dynamic universe in which everything changes and the cycles of birth, life, and death are all about us. At some time, less than 5 billion years hence, the sun will swell to become a red giant and then shed its outer layers to create a planetary nebula that will swallow the earth. Just like everything in the universe, the sun has a finite lifetime. Tens of thousands of years later astronomers on alien planets may watch its planetary nebula and study the radiation in order to better understand the nature of stellar evolution. Those beings will never know who, or what, orbited the original star.

# 5
# Radio Waves from the Milky Way

## 5.1. "A Steady Hiss Type Static of Unknown Origin"

By 1933 Jansky had concluded that the source of the steady hiss he had detected with his antenna must be somewhere outside the earth since it seemed to move through the sky along with the stars in a manner consistent with its being of an extraterrestrial nature. He established an approximate direction for the source as 18 h right ascension and $-10°$ declination. (Right ascension and declination are coordinates astronomers generally use to locate objects in the starry heavens, see Appendix A.7.)

In another report, published in 1935, Jansky stated that the

> ... radiations are received any time the antenna system is directed toward some part of the Milky Way system, the greatest response being obtained when the antenna points toward the center of the system.[1]

Within Jansky's experimental inaccuracies he found the peak radio emission to be located more or less in the constellation of Sagittarius. He attempted an explanation for the mechanism that generated the radio signals, suggesting that stars or interstellar matter might be the cause. We now know that cosmic-ray electrons spiraling about interstellar magnetic field lines produce the bulk of the so-called radio continuum emission from the Milky Way. Jansky also noted that the hissing sound of the radio waves from space was very similar to the hiss produced in his headset connected to the radio receiver.

## 5.2. Receiver Noise—"Listening" to Radio Sources

The reference to "radio" in the term "radio astronomy" sometimes triggers visions of radio astronomers sitting beside a loudspeaker listening to cosmic music. However, it is only fruitful to listen to the sounds emerging from the receivers connected to the radio telescope when trying to identify a source of unwanted

---

[1] K. G. Jansky, *Proceedings of the Institute of Radio Engineers*, Vol. 23, page 1920, 1935.

radio interference. Radio astronomers never actually listen to sounds generated by their radio receivers.

Cosmic radio signals exhibit a characteristic hiss like that in a television when it is set to an unused channel (and disconnected from a cable). This hiss is called noise and is electrical in origin, being produced by random movement of electrons inside the electronic components of the radio or TV set. The noise generated within a television set can also be seen as "snow" on the screen.

Radio waves from cosmic sources are generated by the motion of electrons, either traveling close to the speed of light (relativistically) or much more slowly (nonrelativistically). In either case the random movement of the electrons in radio sources creates electrical (or more correctly, electromagnetic) noise, which is indistinguishable from noise produced by the receivers attached to the radio telescope. Just as it is difficult to hear someone speaking above the noise of a crowded cocktail party, the presence of internally generated receiver noise makes the detection of the distant radio whispers from space difficult.

One of the greatest challenges electronic engineers confront in building radio astronomy equipment is to reduce the noise generated within their electronic components. The task of the construction of low noise receivers has now been elevated to an art, with a considerable fraction of the budget for new radio telescopes set aside for the development of highly specialized low-noise receivers.

The experience of listening to random noise can be extremely soothing to the ear, as in the case of the sounds of distant surf, a bubbling brook, or a waterfall, but listening to the noise from space is of little practical value. The cosmic radio signal needs to be translated into an electrical current, which, in the early days of radio astronomy, was used to drive a pen over a paper chart or converted into data to be handled by a computer for later study.

Radio astronomers, therefore, neither look directly at nor listen to radio sources. Instead, the radio signals from space are processed in computers and displayed in a way that means something to the human eye. At the same time, quantitative information concerning the intensity of the radio signal has to be derived through accurate calibration measurements and calculations of antenna and receiver characteristics. The training of radio astronomers includes learning how to perform these functions and then interpreting what the data signify about events in space.

---

During long midnight observing sessions with the 250-ft radio telescope at Jodrell Bank in the early 1960s we did have a loudspeaker attached to the output of the receiver. This served two purposes. The scientifically respectable one was to detect radar transmissions from Manchester's Ringway airport, which would sometimes leak into the receiver depending on where in the sky the telescope was pointed. If we heard the radar we knew that the data being recorded had been compromised by this unwanted interference. The second reason, for me at least, was to listen for signals from extraterrestrials. After all, why not? I had nothing to loose except that I ran the risk of being lulled to sleep in a midnight observing run because of the gentle sounds of the receiver hiss, known as white noise. Our observations were always in the 1,420 MHz band at which interstellar hydrogen transmits a signal (see Chapter 6). It was widely believed that extraterrestrials would use that band to signal its presence. Needless to say I never heard a squeak from them out there!

## 5.3. Grote Reber Maps the Milky Way

Jansky could hear the faint radio hiss from space in his earphones and went further to report on his quantitative measurements of the intensity of the received emissions. However, his discoveries went largely unrecognized by astronomers, either because they never got to read Jansky's technical papers, which were published in a journal aimed at radio engineers, or because the astronomers, not familiar with radio engineering, simply were not interested. A few people did take note, and it was Grote Reber, resident of the Chicago suburb of Wheaton, Illinois, a radio engineer by profession and a radio amateur (ham), who built the world's first dish-shaped radio telescope (Chapter 2).

In the spring of 1938 Reber set his equipment to receive at a frequency of 3,300 MHz, but he had no success in finding the cosmic static. The frequency was chosen because he expected the radiation from the Milky Way to be thermal in origin, and therefore the sky should be brighter at higher frequencies. But the Milky Way does not emit thermal radiation. It was only realized nearly 20 years later that the actual radiation mechanism is nonthermal (synchrotron emission), and therefore the radio signals are weaker at higher frequencies (Appendix A.4).

Reber decided to try again at a lower frequency and built a new receiver and antenna to operate at a frequency near 1,000 MHz. Again he did not detect any signals. Finally, in his third attempt—he was a patient man and made a large leap in frequency to 160 MHz—he detected a signal from the Milky Way and began a systematic mapping of this "cosmic static." His 31-ft-diameter parabolic dish, which he constructed single-handedly, is shown in Figure 2.1 and an example of his data is shown in Figure 2.2. The resolution of his observations was 12.5°, which may be compared with resolutions of 1 s of arc now commonly achieved.

Reber discovered that greater amounts of radio emission seemed to originate from specific directions, notably Cassiopeia, Cygnus, and Sagittarius, this discovery being the first hint that individual radio sources might exist.

## 5.4. A Radio Map of the Whole Sky

A modern map of what the radio sky "looks" like is shown in Figure 5.1 made at 408 MHz. This map uses galactic coordinates that are defined with respect to the shape of our Galaxy. The bright band associated with what is in essence the Milky Way (known as the galactic plane to astronomers) cuts across the middle of the diagram along galactic latitude 0°. Galactic latitude (b) is then measured north and south to the two galactic poles at latitudes ±90°. Galactic longitude (1)

408   MHz

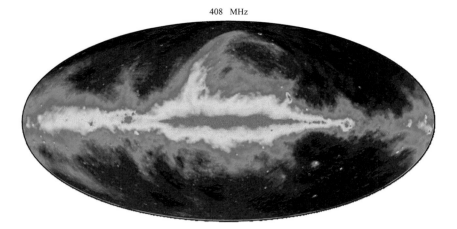

**Judrell-Bank  230-feet  +  Effelsberg 100-m + Parkes 64-m**

FIGURE 5.1. The radio continuum emission from the Milky Way on an all-sky projection. This map was made at a frequency of 408 MHz. (Image courtesy of Patricia Reich, Max Planck Institut für Radioastronomie.)

is measured along the plane of the Milky Way and this map has longitude zero in the center going through 90° to l = 180° at the left-hand end and resumes at the right-hand end. In the galactic plane, emission from a greater depth of the Galaxy is intercepted by the radio telescope. In general the intensity of the received radio waves from those directions is greatest.

[The projection system used to make these maps allows the sphere of the sky we see wrapped all around us to be unwrapped to produce a two-dimensional projection.]

The bright feature sweeping up and out of the galactic plane at the top-center of the map is known as the North Polar Spur. It is a segment of a huge radio emitting shell ejected a few million years ago by a star that exploded within a few hundred light-years of the sun.

Several discrete sources of radio emission are visible as bright points in the maps. The smaller bright dots well away from the plane of the Galaxy are distant, extra-galactic radio sources while those close to the galactic plane are either star forming regions or the remains of stars that have long since exploded to produce radio emitting shells of gas (see the next chapter).

## 5.5.  The Appearance of the Radio Sky

Because of the nature of the nonthermal radiation the radio sky appears brighter as the observing frequency is decreased. At low frequencies such as 100 MHz the entire sky appears to glow, and as the observing frequency is decreased further the

Milky Way band becomes broader and broader. The trend continues until the Milky Way is no longer brighter than the rest of the sky. Instead, at frequencies of many tens of MHz, first the Milky Way and then the entire heavens grow darker where patches of nearby thermal gas between the stars begin to absorb the "background" nonthermal radio signals. The appearance of the invisible radio sky, therefore, depends on the observing frequency.

At no frequency does the radio sky look like the optical sky. None of the few thousand stars we can see at night are radio emitters of any significance. An exception is the central "star" of Orion's sword, which is a nebula visible to the naked eye and is the strong radio source Orion A (Chapter 4).

## 5.6. Polarization of the Galactic Radio Waves

The hypothesis that the broad band of emission from the Milky Way is produced by cosmic rays spiraling around large-scale interstellar magnetic fields is supported by observations of the spectrum (which is a measure of how brightness varies with wavelength) and the polarization of the emission. Relativistic electrons traveling near the speed of light spiral around magnetic fields and in the process produce nonthermal radio emission. This radiation exhibits a property called polarization, which means that the radio waves vibrate in some preferred direction. Imagine a rope held at two ends. Flip one end up and down and a wave travels down the rope. Such a wave is vertically polarized. Hold the rope taut and flip it sideways. A horizontally polarized wave now travels along the rope.

Large-scale surveys of the polarization of the radio emission from radio sources have detected polarization and the same is true for emission from the Milky Way and the North Polar Spur, and these data can be used to infer the angle of polarization at the source. In the case of the Milky Way the magnetic field is largely aligned parallel to the galactic plane. In the Spur it runs along the ridge of the filaments seen in Figure 5.1. In other parts of the sky the field is less ordered.

## 5.7. "Normal" Galaxies

Our Galaxy is believed to be relatively normal, just like billions of others in our universe. We might therefore expect most other galaxies to emit weak radio signals, and that is precisely what is found. However, most of them are so far away, and therefore appear so faint, that they do not show up even in the most sensitive studies of the radio sky. By comparison with radio galaxies and quasars (Chapter 10), normal galaxies are all but invisible to radio astronomers. An exception is M31 in Andromeda, located 2 million light-years away, whose radiograph is shown in Figure 5.2. The center of M31 is a bright source of radio emission and the surrounding ring is due to radiation from star forming regions and the remains of exploded stars. To an observer in M31 the radio map of our Galaxy would probably look very much as does M31 to us.

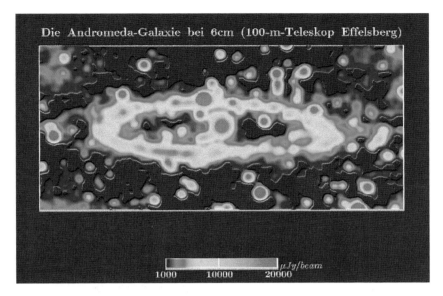

FIGURE 5.2. The radiograph of the Andromeda Galaxy M31 made at a frequency of 5,000 MHz using the 100-m diameter radio telescope at Effelsberg in Germany. The brighter regions, shaded in red, are mostly sources of intense radio waves in M31 such as old supernova remnants, and the center of the Galaxy glows brightly in a manner akin to our own Milky Way. (Image courtesy of Rainer Beck, Max Planck Institut für Radioastronomie.)

## 5.8. A Note on Distances

Distances in astronomy are given in terms of light-years, a convenient unit that makes the numbers manageable. A light-year is the distance a light wave can travel in a year, about 6 trillion miles or 10 trillion km. These numbers are so large we cannot imagine them. Astronomers tend to use parsecs when they talk of distances. A parsec is about 3 light-years.

## 5.9. The Shape of the Milky Way Galaxy

Our Galaxy is shaped like a huge disk, roughly 100,000 light-years in diameter and 1,000 light-years thick, highlighted by spiral swaths of stars and interstellar dust and gas. From our vantage point inside this disk, the stars sweep along a faint glowing band across the sky to create the Milky Way we can see on a dark summer's night in the northern hemisphere. The glow is created by millions of distant stars too far away to be picked out individually. To our eyes the stars that make up the constellations are neighbors, usually between ten and several hundred

light-years away. Far beyond them, hidden from our visual gaze, lurks the dead
center of our Galaxy.

Once every 200 million years the sun, its attendant planets, and all the life
forms we know about move ponderously about that hub located 25,000 light-years
away beyond the stars of Sagittarius. It is there that a huge black hole harboring
the equivalent of 4 million suns within its grasp holds court and dominates the
remarkable activity at the galactic center.

The light from stars near the galactic center is completely obscured from our
view by dust; less than a trillionth of that light can reach us. It requires radio
and infrared observations to penetrate the gloom because their wavelengths are
much larger than the size of the dust particles that drift in great clouds between
the stars. Thus radio and infrared waves travel relatively unhindered while light,
with a wavelength about the same size as interstellar dust particles, is absorbed en
route to earth.

## 5.10. The Center of the Milky Way

When Karl Jansky first discovered the radio signals from space he already concluded that they were concentrated to the Milky Way with the greatest intensity in
the direction of Sagittarius. Grote Reber made measurements with his pioneering
dish-shaped antenna designed to home in on details of the radio sky and confirmed
that the peak of the emission lay in that constellation. In 1959 the central radio
source was recognized as being composed of at least four separate sources, labeled Sgr A, B, B2, and C. Subsequently other structures were found as shown
in Figure 5.3, which covers an area some 680 light-years horizontally (in galactic
longitude) and 53 light-years vertically (in galactic latitude). (This and most of the

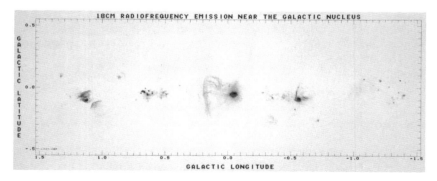

FIGURE 5.3. A panoramic radio view of the inner 680 light-years centered on the core of
the Milky Way Galaxy in Sagittarius, which is otherwise completely hidden from view to
optical telescopes. The vertical dimension of the image corresponds to a linear distance of
53 light-years at the galactic center. Several distinct patches of radio emission described
in more details in the text are, from left to right, known as Sgr D, B2/B1, A (the center),
C, and E. (Image courtesy of Harvey Liszt, NRAO.)

subsequent radiographs in this book were made using data obtained with the Very Large Array radio telescope in New Mexico.)

The whole area around the center of the Galaxy features an amazing variety of radio emitting structures with HII regions created by dense pockets of star formation dominating the emission features. From the left in Figure 5.3 we find Sgr D (longitude 1.1°), an HII region, and supernova remnant close together. It is not surprising that where star formation is active some of those stars will already have exploded as supernovae. Sgr D also contains a cloud of complex molecules (see Chapter 7). Next, around longitude 0.6°, is a double feature known as Sgr B2/B1 which consists of HII regions produced by massive stars situated in the periphery of giant clouds of molecules. Sgr B2 includes the most massive cloud of molecules found to date in the Galaxy.

The center of Figure 5.3 marks the direction of the very heart of our Galaxy, the complex radio source Sgr A at longitude 0°. It is a mix of thermal emission and nonthermal radiation from the great filaments that cut across the galactic plane, defined by the latitude 0° line. Next comes Sgr C, which is littered with shell-like features (not visible in this image) as well as loops and filaments. It too is basically an HII region with the nonthermal filaments looking very much like those in Sgr A (see also below). A giant cloud of molecules is also found in this direction. At the right-hand end of Figure 5.3 is Sgr E, a cluster of small HII regions surrounding massive stars. This cluster contains something like 20 such stars and detailed mapping of the area has revealed as many as 70 discrete "hot spots."

For the observers of the night sky, the strongest radio source marking the galactic center, Sgr A, is situated almost at the boundary between the constellations of Sagittarius and Ophiuchus and lies about 4° beyond the tip of the spout of the "teapot," which defines the visible constellation of Sagittarius. For the serious observer, its location is at right ascension 17 h 42 m 29.3 s and declination −28°59′18″ (1950). This marks the location of a bright, compact, point-like source of radio waves, called Sgr A*, the absolute center of the Galaxy. Until the 1950s it was believed that the center of the Milky Way was located in a direction about 30° away from Sgr A, a misidentification it turned out, hardly unexpected given the difficulty of seeing anything in those directions with optical telescopes except clouds of foreground stars. It was the discovery of this strong radio source that began a trail of research that led to the present definition of where the center of the Galaxy is located.

## 5.11. Close-up Radio View of the Galactic Center

A close-up view of the central 180 light-years of the Galaxy is shown in Figure 5.4. The image is labeled in right ascension (horizontally) and declination (vertically) where the tick marks are separated by 10 arcminutes, equivalent to a linear distance at the galactic center of 75 light-years. The galactic plane cuts across this map from near the top-left side to about one third of the way across the bottom axis. The bright blob at the lower left of the figure marks the Sgr A radio source composed of a very hot gas emitting thermal radiation within which a bright pin-point of nonthermal emission signals marks location of the dead-center source, Sgr A*.

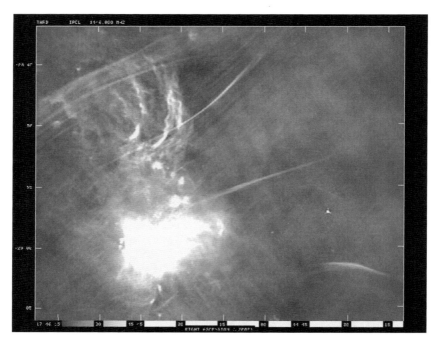

FIGURE 5.4. The central 180 light-years of the Milky Way showing the beautiful complexity of the core of our Galaxy. Within the bright region labeled Sagittarius A, the center of the Milky Way is defined by a massive black hole containing the equivalent of nearly four million times the mass of the sun. Narrow filamentary structures, known as the Northern and Southern Threads, highlight magnetic fields, illuminated by energetic cosmic ray electrons to produce nonthermal radio emission. The more diffuse arched filaments are created by gas heated by two enormous star clusters, the location of one of them, the Arches cluster, indicated by the star symbol. The plane of the Milky Way is oriented roughly along a line from the upper left side of the image through the bright area called Sag A. Investigators: Cornelia Lang, Mark Morris, and Luis Echvarria. (Image courtesy of Cornelia Lang, University of Iowa.) Reproduced by permission of the AAS.

Long magnetically controlled threads of nonthermal emission streak across the region and at the upper left another series of streamers known as the Radio Arc is found. These cross the plane of the Galaxy at right angles. The arched filaments are thermal in nature and are associated with large clusters of very young stars that lie on the surface of another enormous cloud of interstellar molecules. The Arches cluster contains about 100 very massive, young stars and countless smaller ones for a total mass of about 20,000 suns, as is inferred from infrared observations. The arched filaments are located about 85 light-years from the galactic center and can be seen in more detail in Figure 6.3.

At the lower right of the image in Figure 5.4 is another remarkable nonthermal filament called G359.8 + 0.2 (referring to its galactic coordinates).

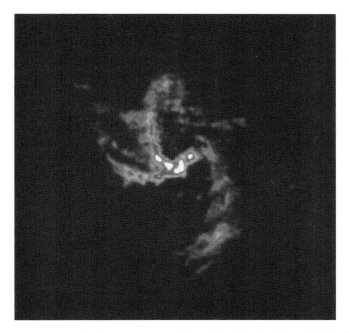

FIGURE 5.5. A radiograph of the very center of the Milky Way showing spiral-like features swirling out from the bright central blob that marks the location of the 4 million solar mass black hole that is at the core of the Sgr A* radio source. To allow the presentation of the data without totally swamping the image with the radiation from Sgr A*, its radio emission has been subtracted from the map. Investigators: K.Y. Lo and M.J. Claussen. (Image courtesy of NRAO/AUI.)

Zooming into the dead center of the Galaxy, Figure 5.5 shows a close-up of the Sgr A where the bright Sgr A* source is located inside an S-shaped or spiral-like region reminiscent of a miniature Galaxy. The material in the clumpy spiral-like structure is mostly thermal, although the compact source, Sgr A*, itself, is non-thermal and typical of compact sources in radio galaxies and quasars (Chapter 10).

Overall, the Sagittarius radio sources represent a stunning and very complicated mix of thermal and nonthermal radiations. The arched filaments and the core seem to be primarily thermal, while the filaments that cross the galactic plane are nonthermal in origin and signify the presence of highly elongated magnetic field lines of force along which a current may be running.

## 5.12. The Very Center and the Black Hole

The galactic center is home to an enormous cloud of molecules with carbon monoxide (CO) as an important tracer of activity in the region. Systematic observations of infrared radiation from stars in a cluster right at the galactic center using the Keck Telescope makes it possible to infer the mass of the object about which they are orbiting. It contains the equivalent mass of 3.7 million suns and its diameter as inferred from radio observations is so tiny that it can only be a black hole that holds those orbiting stars in its grasp. This massive black hole would have a radius of about 10 million km, or about 15 times the radius of the sun. That may seem large, but to cram the mass of millions of stars into so tiny a volume causes one to wonder how that happened in the first place. Has the massive black hole been there since the universe was born, or did it grow as other galaxies collided with the Milky Way billions of years ago to send stars hurtling inward to satisfy the ever more voracious appetite of the growing black hole?

We are lucky to be living in a Galaxy where the effects of the black hole at the center are felt only close to the center, unlike the case for radio galaxies to be discussed in Chapter 10 where explosive events project material outward as far as several times the diameter of the parent galaxies. But of course luck is not involved, because we are only here, now, because our space environment is relatively benign.

# 6
# Interstellar Hydrogen

## 6.1. Clouds of Destiny

Once upon a time, about 300,000 years after the Big Bang that triggered the existence of our universe (Chapter 12), hydrogen atoms were created in great profusion. Hydrogen became the basic building block of galaxies, stars, and nebulae. It is consumed in the furnaces at the cores of stars and converted into helium in the thermonuclear process that generates the energy that keeps the stars shining. At the stellar cores temperatures reach tens of millions of degrees, and it is there that the hydrogen, and subsequently helium, is consumed and converted into heavier elements such as carbon, oxygen, nitrogen, and phosphorus, the building blocks of life that can later be released into space in supernova explosions to become available for planet formation in subsequent generations of starbirth.

The space between the stars in the Galaxy (interstellar space) is filled with diffuse hydrogen where it drifts in its basic neutral (or cold) form known as HI. In Chapter 5 we discussed HII, the ionized form, but it is the swirling masses of HI that tell us about conditions in space before stars form.

During the World War II a small group of astronomers in occupied Holland regularly gathered to discuss topics of scientific interest. It was at one such historic meeting that Henk van de Hulst reported that he had calculated that the neutral hydrogen atom should transmit a detectable radio signal. This meant that this important gas could be directly observed, but not quite yet because no one had a radio telescope or sensitive receiver with which to make a search. However, the Dutch scientists did know that pioneering radio astronomical work had been done in the United States before the war broke out and looming along their coastline were large dishes used as part of the German early warning radar system. As soon as the war ended, one of those dishes would be modified to serve as a radio telescope to search for the hydrogen signature originating between the stars.

## 6.2. Generation of the 21-cm Spectral Line

The neutral hydrogen atom consists of a proton with an electron in an orbit about it. Both the proton and the electron have a property called spin, which can be in the same direction (called parallel spin) or in opposite directions (antiparallel) relative to one another. The total energy contained by the atom in these two conditions is different. When the spin state flips from the parallel condition to the antiparallel, which contains less energy, the atom gets rid of the excess energy by radiating a spectral line at a frequency of 1420.405 MHz, generally known as the 21-cm line referring to its wavelength in the radio band. The 21-cm line is the signature of HI and makes the gas observable to astronomers on earth.

The term "spectral line" refers to the fact that if you adjust the tuning on a radio receiver operating around the hydrogen signal produced by a distant interstellar cloud (containing vast numbers of atoms), the radio message comes bursting through at one frequency. This is similar to a radio station on your AM or FM radio coming through at only one spot on the dial.

When the war ended, a scientific race began using the surplus radar equipment to detect the hydrogen line, and in 1951 three research groups—in Australia, the United States, and Holland—nearly simultaneously—discovered radio emission from interstellar neutral hydrogen.

## 6.3.  Observations of Interstellar Neutral Hydrogen

Within a hydrogen cloud, individual atoms move with respect to each other with the velocity determined by the temperature within the cloud. The spectral signal emitted by individual hydrogen atoms is therefore Doppler- shifted (see Appendix A2.5) to slightly different wavelengths, depending on the motion of the atoms. As a consequence, the so-called spectral line, which would otherwise be observed as a spike coming through on the dial, as it were, is spread over a small range of wavelengths. It is said to be "broadened" by random motions within the cloud. The width of the observed spectral line has long been believed to be a measure of the temperature in the cloud. This means that the radio telescope acts as a giant thermometer capable of probing into the well-being of distant parts of the Galaxy. Cloud temperatures range from 10 K in small, dense clouds (densities hundreds of atoms per cubic centimeter) to 1,000 K or more in large diffuse masses (densities less than one atom per cubic centimeter). These densities are all vastly lower than the air we breathe, whose density is about 10-million-trillion molecules per cubic centimeter.

## 6.4.  An Image of Interstellar Hydrogen

It wasn't until some 50 years after the detection of the HI signal that a comprehensive all-sky survey of the HI spectral line was completed under the guidance of W. Butler Burton at the University of Leiden. He and his collaborators (Dap Hartmann, Peter Kalberla, and Ulrich Mebold) made use of a now decommissioned 25-m-diameter dish at Dwingeloo in the Netherlands, while the southern skies were mapped with a similar-sized dish in Argentina (E. M. Arnal, E. Bajaja,

R. Morras, and W.G.L. Pöppel). The completed project is known as the Leiden–Argentina–Bonn (LAB) survey. To give the reader some feel for the enormous scope of this project, the LAB Survey observed 400,000 directions and obtained a spectrum with 1,000 frequency channels at each location.

> Had I attempted a similar survey back in 1962 with the equipment available to me then, it would have taken 10,000 years to complete the project! This stunning difference is in part due to a factor of 100 improvement in receiver sensitivity, 1,000 data channels as opposed to one, and of the number of directions (400,000) that had to be observed. Sometimes it helps to bide one's time!

Figure 6.1 is an all-sky HI map made from the LAB Survey data where the color is a measure of the total number of hydrogen atoms along the full line-of-sight through the Galaxy in any given direction. The coordinates for this map are again galactic longitude and latitude, just like Figure 5.1. The lines running from left to right are at latitudes $+60°$ and $+30°$ above the plane of the Milky Way and $b = -30°$ and $-60°$ below the plane. Lines of constant longitude are shown running north–south at every $30°$ in longitude.

An intriguing feature of this map is the presence of arcs or filaments (long streamers) visible as great threads of emission, whose shapes are almost certainly controlled by magnetic fields between the stars. Hydrogen gas pushed around by expanding supernova remnants, which have long since faded into oblivion, may be guided along magnetic field structures to produce the patterns seen in Figure 6.1. The hydrogen associated with the North Polar Spur is clearly seen in this map (compare with Figure 5.1). Also seen is a large tongue of HI gas below the galactic plane at $l = 180°$ known as the Orion Spur in which the Orion Nebula (Chapter 5) itself resides, at $l = 209°$, $b = -20°$.

## 6.5. Seeing into the Depths of Space

Motion within interstellar clouds means that the Doppler spread broadens the line. In addition, motion of the entire cloud, either toward or away from the sun, also produces a Doppler shift. For hydrogen clouds moving away from us the radiation is observed at a wavelength slightly longer than expected (redshifted) and for those coming toward us it is shifted to shorter wavelengths (blueshifted). The map of total hydrogen emission in Figure 6.1 added together all the emission at all Doppler shifts pertinent to galactic hydrogen. For directions along the galactic plane ($b = 0°$), the velocity information can be converted into a distance, provided the rotation of the Galaxy is known, and that comes from a study of the motion of stars whose distances are known. Thus the observations of interstellar hydrogen along the galactic plane (the Milky Way band at $b = 0°$ in Figure 6.1) reveals the distribution of the gas in three-dimensional space—two dimensions of position

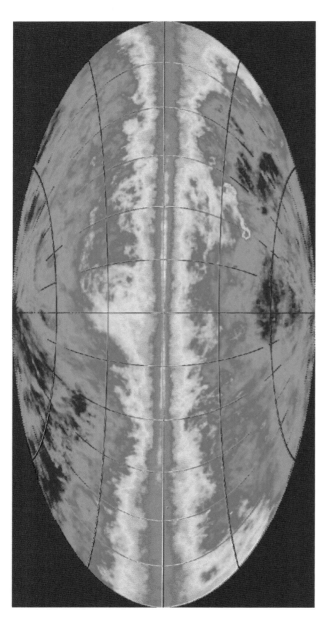

FIGURE 6.1. An all-sky neutral hydrogen (HI) where color corresponds to the total mount of hydrogen found in a given direction. The map coordinates are the same as used in Figure 4.1. The plane of the Galaxy, the Milky Way, follows latitude 0° across the center of the map, and galactic latitude increases to the north galactic pole at the top and the south galactic pole at the bottom. The galactic center at longitude 0° is at the center of this map. An equal areas map like this minimizes distortions that are created in projecting a spherical distribution onto a flat surface. At the right-hand edge an HI tongue to the south merges with the structures seen at the left-hand edge. This is the Orion Spur, about 1,500 light-years distant. A comparison with Figure 4.1 shows the hydrogen gas associated with the North Polar Spur emerging from the Milky Way in the north. The two bright patches in the southern sky well away from the plane of the Galaxy at latitude around 30 and –45°, longitude roughly 280 and 300° are due to HI emission from the large and small Magellanic Clouds, the nearest galaxies to the Milky Way. This version of the HI data was prepared using data from the Leiden—Argentina–Bonn HI all-sky survey data and published in Astronomy and Astrophysics, Volume 440, page 665, 2005. (Image courtesy of Peter Kalberla and Butler Burton.)

(the two coordinates on the sky) and one of velocity (based on the red- or blueshift observed), which gives a rough estimate of the distance.

At this point in most treatises on radio astronomy a diagram is offered to show what the Milky Way Galaxy looks like from some great distance out in space, and such maps are based on the results of the difficult task of interpreting the HI data to obtain distance. The new LAB Survey data have yet to be analyzed in a systematic way to obtain a uniform picture of galactic structure. What is fairly certain already, based on a great deal of other data, is that the sun is located 25,500 light-years from the center of the Galaxy and that the main HI disk is about 300–1,000 light-years thick, with thickness increasing toward the center of the Galaxy.

## 6.6. Anomalous Velocity Hydrogen

Not all is understood about the distribution of HI in the Milky Way. For example, large areas of sky are found to contain HI moving at velocities that are not expected if the gas is confined to the plane of the Galaxy. In particular, when a radio telescope is pointed above or below the galactic plane, only relatively local gas traveling at velocities between ±20 km/s with respect to zero, defined in terms of the average random motion of stars near the sun, should be observed. However, HI at very high negative velocities, which indicates motion toward us, is found at high galactic latitudes. These structures are known as high-velocity clouds, although detailed maps of such features show them to be filamentary instead of cloud-like. Their distance and origin continue to be the subject of controversy. The bulk of these HI structures in the northern sky follow an arc defined by a weak radio shell found in radio surveys such as the one shown in Figure 4.1, a shell believed to be part of an old supernova remnant at a distance of 450 light-years.

## 6.7. Interstellar Magnetic Fields

If an HI cloud is permeated by a magnetic field the motion of a spinning electron is altered minutely. The 21-cm spectral line becomes split into two lines whose difference in frequency is extremely small but it can be measured. In 1959 it was suggested that this phenomenon, known as the Zeeman effect, could be used as a tool for directly determining the strength of the interstellar magnetic field. Based on observations of the polarization of galactic radio waves and of starlight, the theoreticians predicted that the magnetic field strengths of around 10 microGauss should exist. (A microGauss, or a µG, is one millionth of a Gauss, the unit by which these magnetic fields are measured.) In comparison, the earth's magnetic field strength is about a tenth of a Gauss. It is amazing that a radio telescope can be used to measure the strength of a magnetic field at the microGauss level (equal to one hundred-thousandth of the earth's field) in an interstellar hydrogen cloud 10,000 light-years away, which was ultimately achieved.

In May 1968, I used the 140-ft diameter radio telescope of the NRAO (National Radio Astronomy Observatory) in Green Bank to carry out the first ever-successful measurement of the strength of the interstellar magnetic field, specifically in HI clouds in the direction of the Cas A and Tau A radio sources. Fields of 3.5, 10, and 20 μG were measured. This followed nearly 8 years of intermittent attempts to measure the field. I once estimated that by then something like a year's worth of telescope time, world-wide, had been used in unsuccessful attempts by at least four other groups of astronomers to measure the magnetic field strength.

The first result, on Cas A, was only recognized on July 4, 2 months after the data were taken with a new multichannel receiver for which the necessary software had not yet been written to allow me to see what I was observing at the telescope. Twenty hours of observation in Cas A and 60 h of data for Tau A had to be added together to see the magnetic field signatures. In December 1968, all the necessary software was available when I continued my observations and I experienced one of those remarkable highs that come from the thrill of discovery. This time I pointed the telescope at Orion A and after 20 min of observing I saw the magnetic field signature displayed on a video screen. This after endless hours of observing at Jodrell Bank and at the NRAO over an 8 year span of time without seeing anything other than noise. Several more hours of data reinforced the Orion A signal. It took several days to calm down after the elation I experienced that night. Before then no one had suspected the Orion A direction as a likely candidate for a field detection and yet it turned out to exhibit the strongest signal, due to a field of as much 70 μG.

Since then virtually no interstellar magnetic field strength in interstellar HI has been unambiguously measured because the field is inherently weak, probably 2 μG on average, and the experiment is fraught with technical challenges. The same experiment has been done by radio astronomers using other spectral lines to measure magnetic fields in star forming regions of as much as 500 μG and even 1,000 μG.

## 6.8. Neutral Hydrogen in Other Galaxies

Neutral hydrogen exists in abundance throughout the universe and mapping the HI content of relatively nearby galaxies has produced a wonderful series of images, each quite different from the next. Using dozens of radio dishes joined together to simulate a radio telescope many miles in diameter (e.g., the Very Large Array) allows details in those galaxies to be revealed.

Figure 6.2 shows the HI in the spectacular spiral galaxy M81. Here the HI gas follows the spiral arms perfectly, which turns out to be the exception rather than the rule when other galaxies are considered. Very often the HI is seen way beyond the optical image of a galaxy. In fact, HI maps of distant galaxies reveal many pathological cases, or, according to the title of a catalog showing many such examples, a veritable "Rogues Gallery." An example is shown in Figure 6.3, which

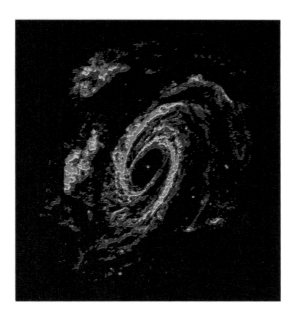

FIGURE 6.2. The distribution of neutral hydrogen gas in the spiral galaxy M81 located 11 million light-years from earth. It is about 50,000 light-years across and is a lone example of the so-called Grand Design spiral with beautifully symmetric spiral arms, which are the exception rather than the rule. In this false color image, red indicates higher gas densities and blue weaker emission. Investigators: D. S. Adler and D. J. Westpfahl. (Image courtesy of NRAO/AUI.)

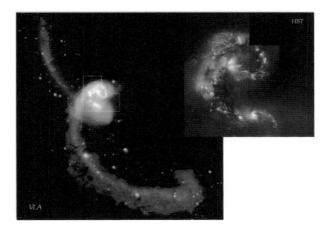

FIGURE 6.3. Composite image of the neutral hydrogen distribution associated with a merging galaxy pair, NGC 4038/9, also known as "The Antennae," superimposed on an optical image of the same area of sky. The two spiral galaxies are in the process of slamming into one another and have thrown off two long, narrow streamers due to past tidal encounters. The hydrogen gas follows the optically visible tails very closely. Investigators: J. E. Hibbard, J. M. van der Hulst, J. E. Barnes, and R. M. Rich. (Image courtesy of NRAO/AUI and J. Hibbard.)

involves a merging pair of galaxies known to optical astronomers as NGC 4038/9 and dubbed "The Antennae" for obvious reasons. As these two galaxies orbited one another, hydrogen gas was dragged out through their close encounters and now follows two arcs that contain stars and interstellar gas.

In some cases HI gas is found between the galaxies, in intergalactic space, and the best example is remarkable ring of gas weaving between galaxies in the Leo cluster.

# 7
# Interstellar Molecules

## 7.1. Chemical Factories in Space

The existence of interstellar molecules is recognized because, just like the hydrogen atom, they emit radio waves in the form of spectral lines. Today over 140 molecular species have been identified in interstellar space. This number compares with 27 in the 1973 edition of this book, and 64 in the 1987 edition. Another 500 or so lines have been observed but not identified. Most of these species are detected at very high frequencies, about 5,000 MHz (short wavelengths, below 6 cm and down to a fraction of a millimeter).

Because conditions in space are vastly different from those on earth, few astronomers 35 years ago would have imagined that water, alcohol, ether, ammonia, carbon monoxide (CO), acetylene, embalming fluid, an amino acid, and even a simple sugar, drift in ethereal clouds between the stars. Even more astonishing is that the vast majority of the identified interstellar molecular species, listed in Table 7.1, are carbon based. The chemistry of these molecules—organic chemistry—is the foundation of all life on earth. It is no longer in the realm of science fiction to speculate that if primitive life emerges on any other planet in the Galaxy, or even in the universe, it will be based on the carbon-based chemical processes (organic chemistry) such as we find on earth and in interstellar space.

Interstellar molecules are usually found either in dense dust clouds or in shells around stars. Two of the most prominent molecular clouds lie in the direction of Sagittarius (in the cloud Sgr B2, which is located near the galactic center), and in the immediate vicinity of the Orion Nebula. Virtually all the complex species in Table 7.1 are found in either one or both of these clouds, not because these are the only two in the Galaxy, but because they are, respectively, the largest known cloud, Sgr B2, and the one closest to us, Orion, approximately 1500 light-years distant.

Thousands of smaller molecule-bearing clouds exist in the Milky Way, with the pervasive observable constituents being carbon monoxide and ammonia ($NH_3$). Formaldehyde ($H_2CO$ a.k.a. embalming fluid) is extremely widespread, while formic acid (HCOOH), the substance that gives ants a remarkably acrid taste, is less common.

TABLE 7.1. Interstellar Molecules

| Chemical Formula | Name | Chemical Formula | Name | Chemical Formula | Name |
|---|---|---|---|---|---|
| AlCl | Aluminum chloride | $CH_3CHO$ | Acetaldehyde | $HCS^+$ | Thioformyl radical Ion |
| AlF | Aluminum fluoride | $CH_3CN$ | Methyl cyanide | $HD$ | Deuterated hydrogen |
| AlNC | Aluminum isocyanide | $CH_3COOH$ | Acetic acid | $HF$ | Hydrogen fluoride |
| $C_2$ | Diatomic carbon | $CH_3NC$ | Methyl isocyanide | $HNC$ | Hydrogen isocyanide |
| $C_2H$ | Ethynyl radical | $CH_3NH_2$ | Methylamine | $HNCCC$ | Ethinylisocyanide |
| $C_2H_2$ | Acetylene | $CH_3OH$ | methanol (wood alcohol) | $HNCO$ | Isocyanic acid |
| $C_2H_4$ | Ethylene | $CH_3SH$ | methyl mercaptan | $HNCS$ | Isothiocyanic acid |
| $C_2H_5OCH_3$ | Ethyl methyl ether | $CH_4$ | Methane | $HNO$ | Nitrosyl radical |
| $C_2O$ | Carbon suboxide | $CN$ | Cyanogen | $HOCH_2CH_2OH$ | Ethylene glycol |
| $C_2S$ | | $CO$ | Carbon monoxide | $HOCH_2CHO$ | Glycoaldehyde |
| $C_3$ | Tricarbon radical | $CO^+$ | Carbon monoxide ion | $HOCO^+$ | Protonated carbon dioxide |
| $C_3N$ | Cyanoethynyl radical | $CO_2^a$ | Carbon dioxide | $KCl$ | Potassium chloride |
| $C_3O$ | Tricarbon onoxide | $CP$ | Phosphorus carbide | $l\text{-}C_3H$ | Propynylidyne |
| $C_3S$ | | $CS$ | Carbon monosulfide | $MgCN$ | Magnesium yanide |
| $C_4$ | Four carbon radical | $FeO^a$ | Iron oxide | $MgNC$ | Magnesium isocyanide |
| $C_4H$ | Butadiynyl radical | $H_2$ | Molecular hydrogen | $N_2^+$ | Dinitrogen ion |
| $C_5$ | | $H_2C_6$ | Hexapentaenylidene | $N_2H^+$ | Diazenylium |
| $C_5N$ | Cyanobutadiynyl | $H_2CCC$ | Propadienylidene | $N_2O$ | Nitrous oxide |
| $C_5O$ | | $H_2CCCC$ | Butatrienylidene | $NaCl$ | Sodium chloride |
| $C_5S$ | | $H_2CCO$ | Ketene | $NaCN$ | Sodium cyanide |
| $1\text{-}C_5H$ | Pentynylidyne radical | $H_2CN$ | Methylene amidogen | $NH$ | Nitrogen hydride |
| $C_6H$ | Hexatriynyl radical | $H_2CO$ | Formaldehyde | $NH_2$ | Aminyl radical |
| $C_6H_6^a$ | Benzene | $H_2COH^+$ | Protonated formaldehyde | $NH_2CHO$ | Formamide |

| Formula | Name |
|---|---|
| $C_7H$ | |
| $C_8H$ | Octatetranyl radical |
| $c\text{-}C_2H_4O$ | Ethylene oxide |
| $c\text{-}C_3H$ | Cyclic propynlidyne |
| $c\text{-}C_3H_2$ | Cyclopropynylidene |
| $CH$ | Methyladine |
| $CH^+$ | Methyladyne |
| $CH_2$ | Methylene |
| $CH_2CHCHO$ | Propenal |
| $CH_2CHCN$ | Vinyl cyanide |
| $CH_2CHOH$ | Vinyl alcohol |
| $CH_2CN$ | Cyanomethyl radical |
| $CH_2D^+$ | |
| $CH_2NH$ | Methanimine |
| $(CH_2OH)_2CO$ | |
| $CH_3$ | Methyl radical |
| $(CH_3)_2CO$ | Acetone |
| $(CH_3)_2O$ | Dimethyl ether |
| $CH_3C_3N$ | Methyl cyanoacetylene |
| $CH_3C_4H$ | Methyl diacetylene |
| $CH_3C_5N$ | 2,4-Hexadiynenitrile |
| $CH_3CCH$ | Methyl acetylene |
| $CH_3CH_2CHO$ | Propanal |
| $CH_3CH_2CN$ | Ethyl cyanide |
| $CH_3CH_2OH$ | Ethanol (ethyl alcohol) |

| Formula | Name |
|---|---|
| $H_2CS$ | Thioformaldehyde |
| $H_2D^+$ | |
| $H_2O$ | Water |
| $H_2S$ | Hydrogen sulfide |
| $H_3^+$ | Protonated hydrogen |
| $H_3O^+$ | Protonated water |
| $HC_{11}N$ | Cyanotetraacetylene |
| $HC_3N$ | Cyanoacetylene |
| $HC_3NH^+$ | Protonated $HC_3N$ |
| $HC_4N$ | |
| $HC_5N$ | Cyanodiacetylene |
| $HC_6N$ | |
| $HC_7N$ | Cyanotriacetylene |
| $HC_9N$ | Cyano-octatetra-yne |
| $HCCCHO$ | Propynal |
| $HCCN$ | |
| $HCCNC$ | Isocyanoacetylene |
| $HCl$ | Hydrogen chloride |
| $HCN$ | Hydrogen cyanide |
| $HCNH^+$ | Protonated hydrogen cyanide |
| $HCO$ | Formyl radical |
| $HCO_2^+$ | Formyl radical ion |
| $HCO^+$ | Formyl radical ion |
| $HCOOCH_3$ | Methyl formate |
| $HCOOH$ | Formic acid |

| Formula | Name |
|---|---|
| $NH_2CN$ | Cyanamide |
| $NH_3$ | Ammonia |
| $NO$ | Nitric oxide |
| $NS$ | Nitrogen sulfide |
| $OCS$ | Carbonyl sulfide |
| $OH$ | Hydroxyl |
| $PN$ | Phosphorus nitride |
| $S_2$ | Diatomic sulfur |
| $SH$ | Sulfur hydride |
| $SiC$ | Silicon carbide |
| $SiC_2$ | Silicon dicarbide |
| $SiC_3$ | Silicon tricarbide |
| $SiC_4$ | |
| $SiCN$ | Silicon cyanide |
| $SiH$ | Silicon hydride |
| $SiH_2^a$ | Silylene |
| $SiH_4$ | Silane |
| $SiN$ | Silicon nitride |
| $SiNC$ | Silicon isocyanide |
| $SiO$ | Silicon monoxide |
| $SiS$ | Silicon sulfide |
| $SO$ | Sulfur monoxide |
| $SO^+$ | Sulfur oxide ion |
| $SO_2$ | Sulfur dioxide |

[a]Uncertain.

## 7.2.  What is a Molecule?

When two hydrogen atoms are brought very close together they lock in an embrace enforced by the cohesive power of their electrons, which orbit both protons. The electrons act as glue that holds the two hydrogen atoms together to form molecular hydrogen. In large molecules, many electrons may be involved in the process of herding (or bonding) together multiple atoms into stable flocks. A molecule can be destroyed, by giving it too much energy. For example, excessive heat or ultraviolet radiation causes the individual atoms to tear themselves loose from their partners. When the adhesive power of the electrons is overcome, the molecule breaks apart to return to its constituent atoms.

In interstellar space there are several ways to form molecules and many ways to destroy them. The fact that molecular species are found in space at all means that the formation process is generally more effective than destruction; otherwise there would be no molecules to be observed! Apparently dust in the interstellar clouds acts to protect the molecules from disruptive stellar energies, especially ultraviolet radiation.

## 7.3.  Molecular Spectral Lines

Most interstellar molecules are asymmetrical in shape, or at least those that have been detected are. For example, an oxygen atom and a hydrogen atom combine in a molecule known as hydroxyl (OH), shaped something like a dumbbell, with the large oxygen and small hydrogen atoms glued together by an encircling electron that originally belonged to the hydrogen. Such a molecule is capable of rotating in two ways; either end-over-end, or around an axis drawn between the two atoms. Whenever two or more energy states are possible, transitions can occur between them and that allows for the emission of energy at specific signature wavelengths, as discussed in the previous chapter for hydrogen. Some molecular species have dozens of possible energy states, depending on their architecture. The signature spectrum of such molecules may contain hundreds of individual "lines" located over a wide range of wavelength, either optical, radio, infrared, or ultraviolet.

The most important interstellar molecule is molecular hydrogen ($H_2$), believed to constitute over 50% of the molecular mass in the Galaxy. Because the molecule is symmetric (being made of two identical hydrogen atoms), it has no differentiated energy states between which transitions leading to radio wavelength spectral lines can occur. Its presence is usually inferred from the widespread distribution of CO, which, so the theory states, is formed only in dust clouds containing a lot of molecular hydrogen. Figure 7.1 shows the distribution of interstellar CO along the Milky Way.

Early in 1985 several of the previously unidentified lines were associated with a ring molecule, $C_3H_2$, and this may be one of the most interesting ring molecules so far detected because ring molecules are so important to life chemistry.

Table 7.1 tends to favor those molecules about which enough is known from lab work or theoretical calculations to allow their identification from spectral

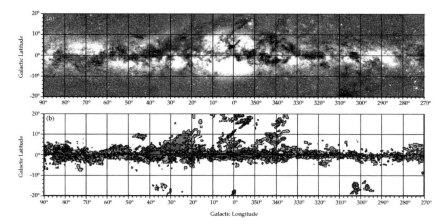

FIGURE 7.1. A dramatic comparison between the visible Milky Way and the distribution of interstellar CO in the lower frame. The coordinate system used here is galactic longitude horizontally and galactic latitude vertically. The galactic center (Chapter 5) is at the center of the map at longitude = 0°, latitude = 0°. The dark dust clouds in the upper frame follow the plane of the Galaxy and also sweep above and below it in various areas. The CO distribution is clearly related to the presence of the dust. The bright galactic bulge at the center is composed of countless stars. The intense CO emission in the direction of the galactic center appears to lie on the bright central bulge but the CO is in fact associated with dust structures too distant to be seen in this image. The optical image is from a panorama produced by Axel Mellinger. Investigators: T. Dame, D. Hartmann, and P. Thaddeus. (Images courtesy of T. Dame and Axel Mellinger.) Reproduced by permission of the AAS.

signatures observed by radio astronomers. They are also those molecules that could be observed from earth because their high frequency radio signals reach the earth without being absorbed in the atmosphere.

A fascinating example of a silicon monoxide (SiO) source is NGC 1333, shown in Figure 7.2. This molecule and a host of others are embedded in a dark cloud of dust and molecules in which stars are about to form. Inside, a massive protostar is ejecting two streams or jets of SiO in what is known as a bipolar flow. One of the jets is headed more in our direction so that the wavelength of its emission is blueshifted as indicated in the figure, the other points away and is redshifted.

## 7.4. Masers in Space

In 1965, while still on the staff at Jodrell Bank, I was fortunate to tour radio astronomy observatories in the United States. After visiting the Massachusetts Institute of Technology (MIT) group and a journey to the National Radio Astronomy Observatory I headed for California. While at UC Berkeley my colleagues there let me into a secret but I had to promise to reveal none of it until their paper announcing the discovery appeared in the journal Nature. What they told me was astonishing. They had been searching for the radio signals from interstellar OH molecules that are clearly distinguished because, unlike hydrogen gas with its single spectral line, OH emits a set of four spectral lines at frequencies of 1612, 1665, 1667, and 1720 MHz. Furthermore, their relative intensities should be in the ratio 1:5:7:1. What they showed me was a very bight line at 1665 MHz with no hint of the other three lines in their data. This made no sense at all. It couldn't be OH; thus they called it "mysterium."

A few days later I attended a meeting of the American Astronomical Society in Ann Arbor where not a word was breathed about this very odd phenomenon. It turned out that radio astronomers from MIT and the National Radio Astronomy Observatory were present at that meeting, and they had also found the signal but no one spoke to anyone else about it because they all realized they were onto something big which they wanted to figure out first.

Upon my return to Jodrell Bank I gave a report of my trip and told my colleagues I could not share the most exciting things I had heard about. A couple of weeks later the report appeared in print, and then I wrote to Berkeley to make a suggestion on how to solve the mystery. I was interested in the search for extraterrestrial intelligence (ET) and what better way for ET to signal its presence than to use one of the four OH lines. Transmission at only one frequency would alert other civilizations to the possible artificial nature of the signal. This theory could be tested if ET also varied the intensity of the signal from hour to hour. They did not reply but a few weeks later they published another report showing that the radio signal did vary with time.

By then, however, others had climbed onto the bandwagon and the mystery was solved. The radio astronomers had discovered a signal that had been amplified in its passage through space, by an interstellar "maser."

The acronym MASER refers to "microwave amplification by stimulated emission of radiation." Microwave amplification refers to the amplification of waves at short radio wavelengths, or microwaves. The better-known acronym, LASER, refers to light amplification by the same process. The stimulated emission of radiation is an interesting phenomenon. Under certain conditions molecules may emit far more energy than expected, provided energy is pumped into them by some external energy source. The OH molecules in clouds around an HII region, for example, can absorb light from very red stars, and that light pumps the molecules into higher energy states and that energy is then radiated away at a preferred radio spectral line frequency. Other interstellar molecules that have been found to exhibit the maser effect include water, silicon monoxide, formaldehyde, and methyl alcohol.

About 200 water maser sources are found in regions of star formation (HII regions). Others are associated with old, highly evolved stars entering their dotage. In either case they appear to be associated with clouds where the densities range from 100,000 particles per cubic centimeter to as high as 100 billion particles per cubic centimeter, about as dense as anything yet discovered in space. To make the

FIGURE 7.2. The bipolar jets emerging from NGC 1333 (gray object near the center of the image) which is a young and bright protostar that will become a sun-like star in a few million years. Located at a distance of 1,000 light-years in the constellation of Perseus, it is embedded in a dust cloud and here the two jets have been mapped in the spectral line silicon monoxide using the Very Large Array. One jet flowing from the protostar is due to gas approaching us and is colored blue. The other jet is receding and is colored red. Gas at the same velocity as the protostar is colored green. Investigator: Minho Choi (KASI). (Image courtesy of NRAO/AUI and Minho Choi, KASI.)

maser work, the pump source has to be about 10,000 times more luminous than the sun. This can be provided by newly formed massive stars or cluster of such stars. The intensity of parts of the maser spectrum, which can be quite complex in shape, may vary from day to day and can be as much as 2,000 Janskies bright. Radio source intensity is measured in units called Janskies (Appendix A.3), named after Karl Jansky, and for comparison the strongest radio source in the sky, Cassiopeia A, logs in at 1,000 Janskies. The brightness of the typical radio sources whose radiographs are shown in this book are in the range from a few hundredths to several Janskies.

It is a little known piece of trivia that during World War II it would have been possible, in principle, to use primitive, first-generation, centimeter wavelength radar receivers to detect interstellar water masers, if anyone had been crazy enough to look for such a signal.

In the direction of HII regions, such as the Orion nebula, the water masers appear in clusters and several small clusters are spread over the area of the nebula. It is believed that each small group of water sources, called a "center of activity," may be due to maser amplification within the envelope, or cocoon, surrounding a specific star. Several such stars lie within the HII region, or very close to its boundary.

Other masers are associated with variable stars that show SiO, water, and OH masers. These stars are known to have molecule-bearing circumstellar envelopes being ejected in a relatively orderly manner, unlike the violent phenomena observed in novae or supernovae. Molecules in these envelopes are pumped to higher energy states by collisions in these gases as the envelopes expand away from the star.

## 7.5. Mega-Masers

A special variety of water and OH masers has been found in distant galaxies. Known as mega-masers, because of their prodigious power, they reveal details of motions of gas in the very heart of galaxies in which star formation activity is high and black holes are in control of things. In these so-called starburst galaxies, observations of the water masers over time can give an estimate of motion close to the black hole at the center. That, together with a measure of the velocity of the "hot spots" in the maser emission, allows the distance to the galaxy to be calculated to within a few percent.

Observations of water masers in the heart of the Milky Way allow a similar calculation to be carried out. The motions of the maser sources around the center of the Milky Way can be calculated very accurately and that leads to a firm estimate of its distance as well as the mass of the black hole, which is close to four million times the mass of the sun.

## 7.6. Giant Molecular Clouds

Giant molecular clouds (GMCs) are the most massive objects (up to 10 million solar masses) in the Galaxy, consisting almost entirely of molecular

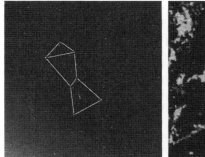

FIGURE 7.3. The Orion CO cloud superimposed on an image showing the location of the primary stars marking the constellation of Orion the Hunter. The Orion Nebula, the center object in the sword of Orion, lies in the densest part of a giant molecular cloud. This cloud is about 1,500 light-years away and contains enough gas to make 200,000 suns. The data used in making this map were obtained with a 1.2-m-diameter radio telescope operated on the roof of the Harvard-Smithsonian Center for Astrophysics, in Cambridge, Massachusetts, which demonstrates that innovative, spectral line research in radio astronomy can still be done in cities. (Image courtesy of T. Dame.)

hydrogen and CO. These are riddled with many of the molecules listed in Table 7.1. GMCs were discovered because they contain enormous quantities of CO, such as the one in Orion shown in Figure 7.3. A GMC, which is usually surrounded by an enveloping cloud of atomic hydrogen gas absent inside the cloud, is typically 150–250 light-years in diameter. The majority of the GMCs, 4,000 of which may exist in the Galaxy, are found between 12,000 and 24,000 light-years from the galactic center. A GMC in Sagittarius is one of the most dramatic objects in the Galaxy and contains 3–5 million solar masses of mostly molecular hydrogen.

The GMCs are stellar nurseries. Evidence for star formation in these clouds comes from the presence of bright infrared sources and masers within the cloud boundaries. The observed HII regions often show hot matter streaming into space, which may be produced by recently formed stars near the edge of the GMCs. These stars then eat their way into the surrounding molecular hydrogen, destroying it and ionizing the atomic (neutral) hydrogen so produced. Ionized gas then streams away from the HII region, which may then appears like a blister at the surface of the GMCs.

## 7.7. The Stages Immediately Following Star Birth

In regions of active star formation, several other phenomena have been observed, again through studying the molecular line emission, which indicate that stars just about to start shining eject a lot of material at hundreds of kilometers per second. These objects are known as T Tauri stars, named after a variable star discovered in the constellation of Taurus. Immediately surrounding a T Tauri star, within an arcminute or so, a small nebula can often be seen. This is made of interstellar dust and gas immediately around the protostellar object. Gas seems to be moving both in and out of the T Tauri star while, incredibly, at some distance further out small nebulae are found streaming away from the star in the same way that a jet in a radio galaxy is pointed away from its nucleus (Chapter 11). These small companion nebulae are known as Herbig–Haro (HH) objects after their discoverers. The HH objects were for a very long time a mystery because they did not appear to have any stellar objects associated with them. The stellar objects are not located in the HH objects, but are light-years away.

It appears that when a T Tauri star, soon ready to turn on its nuclear furnace, begins to stir in its cocoon, the energy it generates, due to the collapse from dimensions of several light-years across to stellar size, is so great that it can hurtle a great amount of matter outward (a millionth of a solar mass per year is typical). This matter travels in two directions because something prevents the material from moving in the other directions. That something is a disk of matter accreting around the star. The accretion disk is shaped like a flat doughnut and gases that are pulled into the accretion disk can escape only by flowing out of the hole in two directions at right angles to the plane of the disk. This is called a bipolar flow.

# 8
# Pulsars

## 8.1. Scintillation of Radio Sources

The pulsar story can be traced back to the mid-1960s, when a pioneering survey in the quest for new radio sources was conducted at Cambridge University in England. Some of the newly discovered sources seemed to change brightness from minute to minute, but only when they were observed close to the direction of the sun. This phenomenon is called scintillation and is produced when the radio waves pass through a patchy cloud of electrons. Such clouds will cause the radio waves to alter their path slightly, jiggling back and forth from minute to minute, which results in the scintillation of the radio source.

The smallest-diameter radio sources scintillate when their beams pass through electron clouds blowing out of the sun in the solar wind. Larger diameter radio sources, however, glow steadily because many beams of radiation from such sources suffer scintillation on the way to the telescope, and when they are added together the average signal is steady. Observation of radio source scintillation thus contains information about both the properties of the particle clouds streaming from the sun and the angular size of the radio sources themselves.

After the discovery of radio source scintillation, the Cambridge radio astronomers realized that a good, cheap radio telescope could be built that would allow persistent monitoring of this phenomenon so that radio source diameters, largely unknown at the time, could be estimated.

## 8.2. The Discovery of Pulsars

Pulsars were discovered because of remarkable persistence on the part of Jocelyn Bell, at the time a graduate student at Cambridge University.

An economy-size radio telescope for studying radio source scintillation had been constructed by eager student labor and it consisted of over a thousand wooden posts each about 10 ft high with miles of wire strung between them. This "telescope" was built before computers were pervasive. Pen recorders were used to display the data by drawing a line on a paper chart that automatically unrolled as the machine created its recording. Analysis of the radio observations required inspection of these charts and the measurement of deflections from the so-called baseline, the normal path of the line drawn on the paper in the absence of a radio source in the beam of the antenna. In modern radio observatories such data are directly fed into computers, where the information is lost from sight until the final numbers are printed out. However, the scintillation experiment produced data on 400 ft of chart paper every day, all of which had to be examined by eye.

The antenna was located near Cambridge and plenty of locally produced radio interference (such as automobile ignition) contributed to the deflections of the pen. These deflections appeared similar to those expected from scintillating radio sources, and Bell, studying her ratio of 400 ft of paper a day, soon became an expert in recognizing which was which. In the course of her work she experienced what other graduate students assigned the tiresome task of studying endless amounts of data sometimes discover—the alert brain is capable of the most extraordinary feats of memory regarding apparent trivia. She noticed that the recordings showed a faint signal that could not be explained by interference or scintillation or any other natural causes then known to astronomers.

Over the months Bell looked at miles of paper charts and found that this "little bit of scruff," as she affectionately named the signal, persisted and that it occurred at night when the sun was below the horizon when no scintillating sources were to be expected. Furthermore, the "scruff" appeared 4 min earlier each day.

The stars move across the skies at a different rate from that of the sun. Another way of stating this is that the length of the solar day, 24 h, is different from the length of the day measured with respect to the stars, which is 23 h and 56 min. All the stars thus appear in slightly different directions, as seen with respect to the horizon, from night to night, an effect that is noticeable to even the casual observer over periods of a month or so. A given star appears directly south, say, 4 min earlier every day.

The "bits of scruff" were unlikely to be human-made interference, which would tend to occur either randomly or at the same time each night. Since the time of arrival of the "scruff," of which she recognized four distinct sources, shifted by about 4 min per day, the signals had to be coming from something associated with the starry heavens. When the research group at Cambridge finally confronted the reality of the discovery, they studied the new radio sources more carefully and were astonished to find that the signals were pulsating with impressive regularity, so regular that it required the best available clocks to measure the arrival time of these "pulse trains." One of the original sources, named CP 1133 (for Cambridge Pulsar at right ascension 11 h 33 min), was found to emit a radio pulse once every 1.33730110168 s. The radio signals were so regular that at first the radio astronomers considered the possibility that they had detected messages from extraterrestrial intelligence (ET), or LGMs, "little green men," as they jokingly called to the first four mystery sources.

The pulsating radio sources turned out to be anything but ET. They are related to an extraordinary object called a neutron star that spins incredibly rapidly and emits radio signals in a beam that sweeps the heavens just as a lighthouse sends its light over the ocean. Every time a neutron star beam swept past the earth a pulse of radio waves was detected.

Pulsars run with clock-like regularity and there is only way to do that in an astronomical context. Some object must be spinning rapidly. Also, each pulsar has its characteristic pulse shape, which refers to the way the radio intensity varies during the pulse. These pulse shapes are the "signature" for each pulsar. From the duration of the pulse, compared to the time between pulses, it is found that the typical pulsar beam is between 10 and 20° wide. All pulsars are also slowing down, some almost imperceptibly yet measurably. This is a natural consequence of aging through the loss of energy by radiation. Some of the really old pulsars even skip beats for minutes at a time.

Because of the very short periods and their relative faintness, pulsars are very difficult to detect. The first few were found because they were fairly strong radio sources and their periods of a few seconds allowed them to be discovered by the radio telescope system designed to observe radio source scintillation.

---

Two enduring memories from the beginning of the pulsar saga have stayed with me ever since. The first relates to a story told me by a colleague from the NRAO who visited us at Jodrell Bank, and then went on to the Cambridge University radio astronomy group. There, while being given a tour of a laboratory by a group of their scientists, he noticed a couple of boxes of computer cards (used in the "old" days to input data to a computer) on a shelf that were labeled LGM 1 and LGM 2. He then returned his attention to his host who was describing some aspect of their research, and when the visitor turned back to take a quick look at the mystery boxes they had been reversed. He was polite enough not to question the work they were doing related to little green men! It turned out they had found the pulsating signals and were not about to talk about it until they had a better idea about what they had found.

My other memory highlighted for me a conviction that theoreticians in any branch of science can probably dream up explanations for any mystery if you only give them an hour or so to think. It doesn't mean they are right, though. In this case in Charlottesville, VA, right after the pulsar discovery had been announced and before anyone had yet figured out what the mystery source of radio pulses might be, a theoretician who shall remain nameless gave a talk about the pulsating radio sources to an attentive audience and listed something like a dozen alternative theories that might account for the weird objects. None of them turned out to be correct! Had one of them been on the mark he could have claimed credit for solving the mystery. Such is life.

---

## 8.3. Where are the Pulsars?

The first step in figuring out what a pulsar is requires finding its distance by making use of a phenomenon known as dispersion. The pulsar signals are modified as they pass through interstellar space, in particular, when they pass through regions of ionized hydrogen between the stars. Interstellar ionized hydrogen—in particular the thermal electrons that coexist with other components in space—causes radio waves to be slightly slowed down compared to the speed of radio waves in a vacuum. A pulse produced at the pulsar will arrive at earth at slightly different times depending on the frequency. This dispersion is measured by observing the arrival times of the same pulse at different frequencies. The amount of dispersion depends on the total number of electrons between the pulsar and earth. Astronomers know the average interstellar electron density is about 0.03 particles per cubic centimeter, and so the distance to a pulsar can be derived.

Most of 1,600 pulsars found to date are located throughout the disk of the Galaxy and tend to concentrate in spiral arms where the majority of stars are also found. Nine pulsars have been found in a nearby galaxy, the Large Magellanic Cloud, but none in other galaxies—so far at least. The farther away they are the weaker the signal is upon arriving at earth.

The majority of pulsar beams do not happen to flash in our direction, and taking this into account suggests there may be 500,000 pulsars in the Galaxy. They should therefore be born at the rate of about one every 10 years, which is odd, because supernovae, the likely source of pulsars, appear to be born at the rate of about one every 50–150 years.

## 8.4.  Formation of Neutron Stars

In 1968 a pulsar was discovered inside the Crab nebula (Figure 8.1) and the pulsar has also been detected optically and through X rays and gamma rays. It flashes at a rate of about 33 pulses per second, and its existence is confirmation of the theory that pulsars are created in exploding stars.

Whatever the nature of the pulsar itself, it has to be spinning extremely fast. Normal stars would shatter long before they could spin as fast as pulsars. The only form of matter that is capable of accounting for the pulsar behavior is a star consisting entirely of neutrons.

Creation of a neutron star involves the catastrophic collapse of the core of a fairly normal star of at least four solar masses whose outer layers explode in a supernova. The explosion is the consequence of a sequence of events triggered when the core of the star runs out of fuel. At that stage the internal fire, which kept the star shining, is extinguished. Until then it was the internally generated heat that balanced the inward pull of gravity to keep the stars stable. When the fire dies out, the core cools, gravity suddenly dominates, and the core collapses. In this collapse, fundamental particles of matter—protons and electrons—are driven so close together that they fuse and become neutrons. As a result, a solid ball of neutrons is produced at the center of the star.

In some cases the core collapse continues with such violence that the neutrons are forced even closer together and, in turn, swallow each other in their own gravitational pull. This is how a black hole is formed. The existence of a black

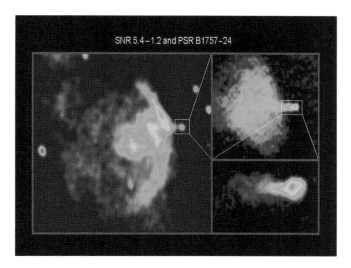

FIGURE 8.1. The supernova remnant G5.4-1.2 (large scale image) called the "Duck" Nebula because of it peculiar shape. The bright blob at the head of the duck is at the location of the pulsar B1757-24 shown in the bottom-right inset. The pulsar has traveled beyond the boundary of the shell of the supernova remnant, which lies about 15,000 light-years away in the constellation of Sagittarius. The pulsar's motion through space across the nebula has been measured from its change of position over several years as about 600 km/s. Investigators: Bryan Gaensler and Dale Frail. (Image courtesy of NRAO/AUI.)

hole can be recognized when it is in an orbit about a companion star. The black hole may draw matter out of that star much like a vacuum cleaner sucks dust from some distance away. The gas plummets toward the black hole and spirals inward into an ever-decreasing orbit to form an accretion disk. The gas in the disk heats up and radiates intense X rays (observed from earth) en route to disappearing into the black hole itself.

Returning to the machinations in the supernova event, when the core has collapsed to form a neutron star, layers of gas above the neutron ball suddenly find that there is nothing to hold them up against gravity. Momentarily they hang suspended and then crash downward, smashing onto the neutron mass and rebound in a violent and fiery explosion. We may see the resulting stellar cataclysm lighting up our sky.

The newly born neutron star will be spinning extremely rapidly as a natural consequence of its having contracted so much. This is due to the conservation of angular momentum, also displayed by a spinning ice skater who begins a spin with arms outstretched and then spins faster and faster as he draws his arms in. This action changes the effective radius of his spinning body, determined by how far his arms are outstretched. A diver doing somersaults uses the same principle by tucking her body in at the start of the somersaults, causing her to become a smaller object, which rotates faster. When she stretches out just before entering the water the somersaults are slowed to a near stop. In the case of the spinning neutron star, shrunk to some small size, its gravitational pull remains sufficiently large to hold the neutron ball together against the disruptive force of rotation.

Thousands of years later the ejecta from the exploded star created wonderful nebulae, such as the one seen in Figure 8.1, "The Duck." This is a radiograph of the supernova remnant G5.4-1.2 (referring to its coordinates in the galactic longitude, 5.°4, and latitude, −1.°2) in Sagittarius and located 15,000 light-years away. The associated pulsar is B1757-24 and its immediate surroundings are shown in the insets. Measurements of the change in the pulsar position over a period of 6 years show it to be traveling at 600 km/s through space. Such a high velocity implies that it must have been torn free of a companion star at birth. The pulsar has escaped the bulk of the nebula and has dragged a trail of radio luminous material along with it.

## 8.5.  Binary Pulsars—Nature's Fabulous Space Labs

During 1874 a major pulsar search was launched at Arecibo observatory. Joe Taylor and Russell Hulse devised an elegant technique that allowed them to discriminate against interference and quickly recognize a pulsar. They found 40 new candidates. One of these, in the constellation Aquilla and labeled PSR 1813+16, turned out to be very peculiar, even for pulsars. The pulses occurred on average every 0.05803000 s, but this rate was not constant, unlike all the other pulsars observed before. Its period showed a 7-h 45-min cyclical change. PSR1813+16 appeared to be binary pulsar, a neutron star in orbit about another object. Pulse rate changes were produced by the Doppler effect, which caused the arrival time of pulses to speed up or slow down (by 16.84 pulses/s with respect to the average) as the pulsar moved either toward or away from the earth during its orbit.

Since at least half the stars in the Galaxy are locked in binaries, a binary pulsar should have come as no surprise. However, the nature of this binary was extraordinary. Careful timing observations enabled the variations in the pulse arrival time to be interpreted with sufficient accuracy to allow the precise orbits and the masses of the pair of stars to be estimated. PSR 1813+16 consists of two objects, each of about 1.4 solar masses, traveling around each other at hundreds of kilometers per second in orbits so close that the distance between them ranges from 1.1 to 4.8 times the radius of the sun (which is about 650,000 km). The maximum diameter of the pulsar's orbit is only a million kilometers.

The binary pulsar provided a fabulous additional bonus. It is a perfect clock in orbit about a massive object, the ideal laboratory for testing Einstein's general theory of relativity. In 1915 Einstein had developed an elegant way to describe gravity and its effects and had explained an observation made during the previous century, that Mercury's orbit about the sun shows an anomaly not accounted for by other theories. Mercury's point of closest approach to the sun, known as its perihelion, moves slowly around the sun at a rate of 43 arcseconds per century. This is called the "precession" of the perihelion. Einstein's theory explained this phenomenon and now the discovery of the binary pulsar provided a further test.

The pulsar is in an elliptical orbit about another object and their point of closest approach (known as the periastron) also precesses. Changes in the pulse arrival times (due to the Doppler shift) should show tiny variations as the pulsar's orbit itself slowly swings around in space. The effect was measured to be 4° per year, precisely as predicted by relativity theory.

The binary pulsar, however, turned out to offer an even more exciting prospect for the radio astronomers, an opportunity unique in the history of the science. The pulsar presented a novel way to test one of Einstein's most important predictions, that objects accelerating in a strong gravitational field should emit a form of radiation called gravitational waves. Just as radio waves are produced by accelerating electrons, gravity waves should be produced by accelerating matter. In the case of the binary pulsar the conditions for radiating gravitational waves appeared to be perfect. Two massive objects moving around one another are constantly accelerating within each other's gravitational influence.

Einstein had stated that he believed gravitational waves would never be detected on earth because they are far too feeble to produce any measurable effects. Despite his caution, however, several laboratories have, with a notorious lack of success, attempted to detect gravitational waves. Now the radio astronomers realized they could search for the effect on the pulsar orbit as a consequence of the radiation of gravitational waves. They would not directly search for the waves, but see what happened to the binary orbit as the system lost energy in the form of gravitational radiation. The energy loss should be manifested as a very small change in the orbital period. This is the consequence of conservation of angular momentum, discussed before. As the system loses energy its orbit shrinks; the pulsar will move a little faster through space, hence the time taken to complete one orbit decreases.

Six years later, after extensive monitoring of the radio pulses from the invisible object in Aquilla, the pulsar orbital period was found to be slowing down by $6.7 \times 10^{-8}$ s/year, equivalent to a shrinkage of 3.1 mm/orbit or 3.5 m/year. This was just the amount that should result from the radiation of gravitational waves. This remarkable measurement, confirming a prediction of a theory proposed 66 years earlier, has proven to be one of the most exciting bonuses produced by radio astronomy research. In 1983, Taylor and Hulse were awarded the Nobel Prize in Physics for this discovery.

But why are these two objects, the pulsar and its invisible companion, so close together in space? The other object is likely to be a neutron star, perhaps a pulsar, but its beam of radio waves does not happen to sweep past the earth. The two objects could not have been so close when they were normal stars. The explanation runs something as follows. Once these were two normal albeit quite massive members of a binary star system. The more massive one evolved quickly, consumed its fuel, and died in a violent supernova explosion. The neutron star stayed in orbit about the other star, which, in turn, reached old age and began to expand to form a red giant. The neutron star then became enveloped in the red giant's atmosphere, where it experienced frictional drag, slowed down, and slowly spiraled deeper into the giant star. The neutron star would not suffer undue hardship at this point, but this would produce severe reactions in the giant star. In due course the star exploded and produced the second neutron core. Today the two neutron stars are in close orbit in the binary pulsar PSR 1813+16, nature's most remarkable laboratory in space.

## 8.6. Millisecond Pulsars

The discovery of the first millisecond pulsar is another tale involving persistence and following clues that could so easily have been ignored. This story also begins with the scintillation of small-diameter radio sources. In addition to the effect described before, the apparent angular size of a radio source is made to appear larger when the radio waves pass through clouds of electrons in interstellar space to produce scintillation, which is most noticeable at low frequencies. Because of this effect, no distant radio source located in a direction close to the plane of the Milky Way would show scintillation because interstellar scattering causes blurring. In such a case a point-like radio source produces a disk-like image, and it will not scintillate.

The clue that led to the discovery of the millisecond pulsar, which flashes at a rate close to a thousand times per second, was a mysterious entry into a catalog of radio sources which indicated that a scintillating radio source, called 4C 21.53, was located close to the galactic plane. According to the theory of interstellar scattering, this was not possible. Early pulsar searches showed no pulsar at the position of 4C 21.53, so the reason for its scintillation was a mystery.

During subsequent research it was discovered that due to a rare error in the original survey, two radio sources were masquerading as one. This still did not explain why the source was apparently scintillating, but it did focus attention on discovering what the radio source looked like at high resolution. These observations revealed that a tiny source appeared to be located next to a larger one. Furthermore, continued observations of the source did not always reveal the mystery scintillation effect. It turned out that the pulsar that did exist there was not readily discovered because its pulses were affected by interstellar scintillation. This caused the pulses to remain hidden for minutes at a time. Since no one expected pulses at the rate at which this neutron star was transmitting, the discovery was made even more difficult. The confusion was sorted out by the persistent work of Donald Backer at the University of California, in Berkeley, and a team of collaborators in the United States and Europe. They opened their minds to the possibility that an extremely rapid pulsar was involved. That led to the discovery of the so-called millisecond pulsar, PSR 1837+21, located 16,000 light-years away in the constellation Vulpecula, which flashes at the rate of once every 0.0015578064488724 s.

The pulse frequency is about 642 Hz, or E above high C on the piano. Audio recordings of PSR 1837+21 made at the 1000-ft-diameter Arecibo radio telescope allows the high-pitched humming sound of the pulsar to be heard quite clearly. Its staggering pulse rate is faster than any other pulsar, and according to conventional pulsar theory a neutron star spinning this fast should be very young—but no supernova remnant was located at its position.

Young pulsars are expected to run down quickly, yet the period of PSR 1837+21 was nearly perfectly constant, which implied that it was very old. So it appeared to be both young and old! Why this apparent contradiction? The answer is that the millisecond pulsar is believed to be a recycled pulsar! It must once have been a normal pulsar, a member of a binary system. Then, as its companion aged and swelled in size toward the end of its life, the neutron star may have gobbled up the companion. The two may literally have blended and in the process the old pulsar's spin rate was greatly speeded up, because in the process of absorbing matter it had to spin faster in order to conserve angular momentum.

In general, millisecond pulsars are all regarded as recycled pulsars. Their very rapid spin rate has been caused by the addition of matter onto an otherwise old pulsar.

## 8.7. What Pulse Timing Tells Us?

The millisecond pulsar provides astronomers with the most accurate clock in the universe. It is not subjected to the complex variations observed in the binary pulsar timing experiments. This clock is so accurate that the millisecond pulsar allows a whole host of other phenomena to be explored. For example, the delay in pulse arrival times due to the gravitational effect of the sun on the radio wave as it travels through space, and the changing gravitational field in the vicinity of the earth as it moves in a slightly elliptical orbit about the sun, are potentially detectable. Time dilation effects (the way a clock appears to slow down as it travels faster) due to changes in the earth's orbital velocity have already been measured and support Einstein's prediction of this effect. Through continued observations of millisecond pulsars it will become possible to discover whether gravitational waves are sweeping into the earth, causing it to "shudder" ever so slightly. Something similar will be found if powerful gravity waves should smash into distant pulsars. Such waves may cause an otherwise imperceptible wobble in the earth's or the pulsar's motion, and a disturbance as small as is expected from this "buffeting" may be revealed by long-term observation of millisecond pulsars.

A total of 150 millisecond pulsars have been discovered and the race is on to use them to detect gravitational waves. The trick is to use several candidates spread around the sky whose pulse rate and change in pulse rate are measured to very high accuracy. Together they form a frame of reference with respect to which the slightest change in the earth's location will be detected. It is not as if the gravitational wave will throw the earth off course. It will suffer a minute judder at best and comparison of the way the pulse trains from the selected millisecond pulsars are affected will, in principle, allow an estimate of the direction in which the gravitational wave was traveling.

It is believed that space around us is filled with gravitational waves at any time but they are infinitesimally weak. The goal of the pulsar experiment will be to detect ones that stand out above the background. A slightly more intense gravitational wave may be created if a distant star should collapse to form a black hole. That event will trigger a gravitational wave pulse that would stand out above the background when it reaches the earth.

The pulsar detection and monitoring equipment used in these experiments boggles the imagination. The search involves extensive computer-aided searching of enormous amounts of radio data in order to recognize the pulsar. In one pulsar search system about 100 Gigabytes of data are collected every hour. Analysis of a 7-h observation takes several years of CPU time, and a cluster of 16 processors is employed to preprocess the information. Even then, collaborators on the search projects employ as many as 100 computers to find hints of these, the most elusive of pulsars.

## 8.8. Pulsars in Globular Clusters

One of the "hot" topics in pulsar research is the study of those found in globular clusters, tight groupings of old stars that orbit the center of the Galaxy but outside the disk of the Milky Way. Some 120 millisecond pulsars have been found in globular clusters, with a preponderance in two clusters, Terzan 5 and 47 Tucanae. And the search for more goes on.

The cluster pulsars are concentrated toward the cores of the clusters, which have long since lost all their interstellar matter. It is there where stars come close to one another and many are locked in binary orbits. Yet there a large proportion of single pulsars exist in the clusters, and there is some indication that a pulsar wind, created by its intense radiation, has blown the gas off a possible companion, essentially evaporating that star.

# 9
# The Galactic Superstars

## 9.1. The Curious Object SS433

When SS433 was first noticed in the 1970s it appeared to be no more than a faint red star, except that it showed hydrogen emission lines. These are spectral lines generated by hot hydrogen at the surface of the star, an interesting phenomenon to two astronomers, C. Bruce Stephenson and Nicholas Sanduleak. This star was the 433rd entry in their catalog of such objects.

In 1976 X rays were discovered to be coming from the direction of SS433 and then, in 1977, radio waves were observed from the same position. This sounded an alarm in the minds of many astronomers. A star that emits unusual amounts of both X rays and radio waves deserved a closer look. What was revealed stunned the astronomical community. SS433 is a small-scale version of the phenomenon that powers radio galaxies and quasars (see next chapters).

SS433 is located 18,000 light-years away inside an old supernova remnant in Aquila first observed in the late 1950s and known as W50. Its radio portrait is shown in Figure 9.1. The remnant is believed to be about 40,000 years old, and its size has swollen to 200 light-years across, enveloping hundreds of stars in the process.

In 1978 routine studies of the spectral lines emitted by this peculiar star were begun in order to see if these would give a clue as to why the object emitted such strong radio waves and X rays. The first detailed observations were so startling that the astronomers involved thought that something had gone wrong with their equipment! This single "star" showed three sets of spectral lines, quite unprecedented in astronomy. One set was apparently normal and showed a small Doppler shift of 70 km/s, expected for the star's direction and distance in the Galaxy. This indicated that the star was partaking of relatively normal motion around the center of the Galaxy in concert with neighboring stars in its vicinity. However, the other two sets of spectral lines were bizarre. One set indicated an extraordinarily high redshift, indicating motion away from the earth at 50,000 km/s. If this were to be interpreted as a typical redshift (as is observed in distant galaxies) the object would have to be over one billion light-years away, hardly a star in our Galaxy! The other set of lines showed a blueshift (motion toward us) of 30,000 km/s. The

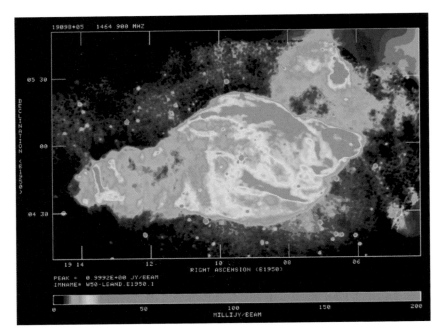

FIGURE 9.1. This is a radiograph of the supernova remnant W50, which houses the highly unusual compact object SS433 near its center. The image was made at a frequency near 1,400 MHz. Investigators: Gloria Dubner, Mark Holdaway, Miller Goss, and Felix Mirabel. (Image courtesy of NRAO/AUI.)

star appeared to be both moving away and toward us, at some sizeable fraction of the speed of light, even though it simultaneously appeared to be moving normally! But this very peculiar beast had more shocks in store for the astronomers, who began to flock to their telescopes by the dozens to observe this cosmic wonder.

Repeated observations revealed that the odd spectral lines were not constant in time. They showed an amazingly regular change in their Doppler shift, which implied a systematic change in velocity. One set of lines varied between a redshift of 50,000 to 0 km/s while the other set varied between a blueshift of 30,000 km/s and a redshift of 20,000 km/s. The cycle repeated every 164 days. The average velocity for the spectral lines was about 12,000 km/s. Why wasn't it close to zero, the velocity of the star itself? After all, the third set of spectral lines showed only a small Doppler shift of 70 km/s due to the star partaking of normal galactic rotation.

The stunning stellar fireworks display generated by SS433 could be explained if it were ejecting two jets of luminous material. Astronomers were suddenly confronting the fact that a quasar-like object (Chapter 11) exists in our galactic backyard. Also, it had to be powered by some remarkable central engine. Because the gas in these jets was moving so fast, another important discovery was made. According to relativity theory, if the jets are produced by material moving at a significant fraction of the speed of light across the sky, and not just either away or toward us, a second phenomenon comes into play. This is the so-called transverse Doppler effect, which is a shift in the spectral lines due to motion across our line of sight to the moving object. There is no everyday analog of this phenomenon, which only becomes important when the object travels at a significant fraction of the speed of light. The variation in the spectral lines due to the jets could be explained if material in the jets were streaming at about 75,000 km/s, an incredible 25% of the speed of light. That would produce a transverse Doppler shift of 12,000 km/s, the average observed velocity for the spectral lines.

## 9.2. A Black Hole and its Accretion Disk

In SS433 two jets appear to be blasting away from some central compact object. A relatively normal star is in orbit about it, a conclusion founded on further observations, which showed periodic 13-day changes in the spectral line velocities from SS433. The source of the jets is therefore a member of a binary star system. But what causes the 164-day period? It cannot be blamed on a neutron star, because they spin at the rate of once per second. Could it be a black hole, an object so massive and so tightly compacted that matter has collapsed in on itself to the point that gravity prevents anything from escaping, even light. A black hole would spin even faster (if its spin could be detected). On the other hand, a normal star rotating once every 164 days would not expel jets with such violence.

After several years of study the likely explanation for the SS433 system emerged: a normal star is in orbit about a black hole that contains the equivalent mass of about four suns. Due to the close proximity of the nearby star, the black hole lures material from the surface of the star and draws it toward the black hole. As the infalling material gains speed, it begins to accrete into a disk of matter that spins around the black hole en route to its final destination—nothingness. In the accretion disk the density grows larger and larger the closer to the black hole the matter comes. In so doing, the particles undergo increasingly violent collisions with each other. However, there is a wonderful twist to this story.

If too much material rushes into this accretion disk a condition known as supercritical accretion is reached and then things become very interesting indeed. An enormous increase in particle collisions suddenly heats the gas to the point where it contains so much energy that it explodes, driving material outward again, to escape the impending clutches of the black hole. But this material cannot blast through the surrounding material in the disk. It can only escape up the central hole of the doughnut-shaped accretion disk. So away we go; two jets blasting outward at a quarter of the speed of light, a very chaotic state of affairs.

The two jets tear into space, gathering up more material as they go. They also expand sideways at about 2,000 km/s and fan out slowly as they rush into the surrounding supernova remnant. The jets themselves appear to be something like long, miniature cylindrical supernova remnants! The hot material in the jet also emits X rays, as observed by X-ray astronomy satellites in 1976.

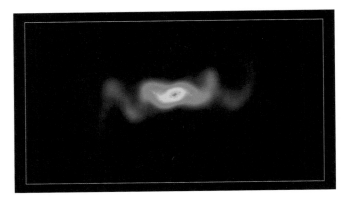

FIGURE 9.2. A beautiful radio image of the galactic microquasar SS433, which is located inside the supernova remnant W50 seen in Figure 9.1, showing the corkscrew motion of the material ejected from the vicinity of the black hole at its center. Investigators: K. Blundell and M. Bowler (Oxford). (Image courtesy of Katherine Blundell.) Reproduced by permission of the AAS.

According to this picture, SS433 should show two nice straight jets pointed away from the central source, in which the velocity of material streaming outward would remain constant with time. But they don't look like that. The radio observations (Figure 9.2) show that the jets are shaped like corkscrews whose twisting motion can be followed from day to day. The reason for this cosmic corkscrew is related to the 164-day period in the jet velocities.

## 9.3. Precession of the Accretion Disk

To explain the twisted jets, the accretion disk appears to be wobbling about the black hole. But why? Because the binary star companion of SS433 is feeding the voracious black hole an excessive diet of gas, mostly hydrogen, and at the same time pulling on the accretion disk. That should be enough to cause slow precession, but there is more to it. The nearby star is not round! Due to the proximity of the black hole it is distorted, and therefore the gravitational influence from the rotating, distorted star is very nonuniform. Note that the accretion disk surrounding this four-solar-mass black hole is only solar-system-sized and the black hole a few kilometers in diameter. (All these numbers and details of the picture described here come from a tremendous amount of research by dozens of astronomers studying the spectral line shifts and a comparison of all the observations made at optical, radio, and X-ray wavelengths, which are then compared with theoretical calculations.)

The result of the tug-of-war between the ugly star and the accretion disk is precession, which is like the wobble of a top set spinning on a table. However, the entire disk does not really move as a solid object, because the gas is passing through the disk so rapidly that today's accretion disk is almost a new one compared to yesterday's.

Now the particles in two jets are ejected straight out into space, but because the orientation of the disk changes with time the direction of ejection also changes. Even as individual particles head straight outward, they create the corkscrew-like pattern of radio emission seen in Figure 9.2. Their trajectories may be likened to water streaming out of a rotating garden sprinkler. Each water drop heads straight out, but as the sprinkler head spins it creates an apparent spiral of ejected liquid.

Seen from earth, the velocity of material in the SS433 jets cycles through a range of values determined by the geometry of the twisted jets with respect to our point of view. Sometimes a jet would point more directly toward us, and days later it would be tilted away from us. This happens with the precession period as the accretion disk, once in 164 days.

To summarize, the strange spectral lines from SS433 are produced by two jets of incandescent gas driven out of an accretion disk surrounding a black hole, which is in orbit about a star that supplies the fuel! The jets are driven explosively outward by the energy created in supercritical accretion, which occurs when too much gas is made available for the black hole to swallow in one gulp. The jets, in turn, are propelled to the outskirts of the surrounding supernova remnant, which they keep fed with energy that makes the remnant shine.

This type of object is now known as a microquasar. The radio observations have led to a determination of the distance to SS433. Since the velocity of material along the jets is known, and the movement across the sky can be seen in the radio maps, the distance to the object can been determined—18,000 light-years, the distance to W50. The light from this remarkable object has been traveling since Homo sapiens dwelt in caves in the last ice age, when humans were utterly oblivious of the remarkable cosmic wonders that exist beyond the stars overhead. Such is progress!

The study of the microquasar that looked like an apparently innocuous little star called SS433 gave astronomers the first clear insight into the physical processes occurring near black holes. SS433 is still being thoroughly studied and the picture to account for its behavior is about as complete as any in astronomy, which is all too often spiced with mysteries which cannot be solved with present day observations and always seem to require bigger telescopes. SS433 is also a wonderful manifestation of the phenomenon occurring in radio galaxies and quasars (the next chapter), but here it is on a tiny scale, very close to home. Its discovery has reinforced the notion that jets and precessing accretion disks are enormously widespread in the universe.

## 9.4. Radio Stars

The pathological object SS433 is certainly the strangest astronomical phenomenon ever observed in our Milky Way, but what about its approximately 250 billion other stars? Each star is expected to emit radio signals by the thermal emission process for no other reason than that the star's surface is hot, at a temperature somewhere between a few thousand and a few tens of thousands of degrees Kelvin. Very hot stellar atmospheres, up to several millions of degrees, are also common. However, the radio emission from the majority of stars like the sun is not detectable on earth because those "normal" stars are too far away, hence their signals are too faint to be detected. If the sun were placed at a distance of 4 or 5 light-years, the distance of the next nearest stars, its radio signals would barely register a flicker with the world's largest radio telescopes. Nevertheless, there are several categories of stars, known as radio stars, which do emit radio waves detectable at earth, and each category does so for different reasons.

The names of the various classes of radio stars are as colorful as the variety of phenomena involved. Irregular and infrequent radio blasts are emitted by RS Canis Venatici (RS CVn), Algol-type binaries, M supergiants, UV Ceti-type flare stars, AM Herculis stars, symbiotic stars, novae, and VV Cephei stars. Each generates peculiarly intense radio signals. Many radio stars generate thermal radio emission in strong winds of gas blowing out of the star or in ejected spherical envelopes of material expanding away from the star's surface.

Some stars appear to have atmospheres as hot as 9 million K, as, for example, the variable star UV Ceti. This is a single star, with no binary companions, and was one of the first stars to be seen to exhibit giant flares, which generate intense nonthermal radio emission. These flares are far more violent than any flare on the sun. During a flare the light from UV Ceti can increase in brightness within minutes, before dying away again. Such a flare appears to be similar to a solar flare and it became a great challenge to pick up its radio signals, even in the days before the technology really allowed a successful experiment. The problem was partly due to technological limitations, but also to the incredible faintness of the radio burst, which would have been all but indistinguishable from ground-based electrical interference and so would have made the observations highly suspect. Only when the large radio interferometers of the late 1970s and 1980s began operation did flare star observations become reliable.

Flare stars are common amongst T Tauri stars (see Chapter 8), which vary in brightness and occur in intimate association with interstellar molecular clouds. These stars are believed to be very young, between 900,000 and one million years old. They may be surrounded by accretion disks and seem to be blasting out matter in the form of minor jets which drive outward and push up against surrounding interstellar matter where diffuse nebulosity may be produced.

It is now believed that all stars may pass through this flaring phase in their early childhood. The optical flaring is easy to see and even small telescopes can detect this activity.

---

The quest to detect radio waves from flare stars was tackled by Bernard Lovell, Director of the Jodrell Bank radio observatory (a.k.a Nuffield Radio Astronomy Laboratories) in the early 1960s using the giant 250-ft radio telescope.

In retrospect he was ahead of his time because the technology for success was not yet available. Lovell expected that radio waves from optical detected flare stars would show a rapid rise and then fall again within minutes. This posed a problem because passing trains on a nearby, electrified track were capable of producing just that sort of signal, a.k.a interference. The British rail authorities did work closely with Lovell on the maintenance of their electrified lines so that sparking would be cut to a minimum. But it was the threat of a new subdivision to be constructed about 15 miles away that had Lovell concerned. The additional automobile traffic would produce added radio interference that would make the flare star experiment even more difficult.

All of the automobiles of the staff and students (those who could afford a vehicle) were equipped with suppressors attached to each spark plug lead so as to cut back on radio signals radiated by the ignition system. To bolster his case with the local authorities to deny the request for a new subdivision, Lovell needed to prove that the added traffic would reduce the ability of the world's largest radio telescope to function effectively. He asked my friend and colleague, Pat Wild, who owned an ancient jalopy, to drive his car out along certain distant roads, after removing the suppressors from his ignition system. Then I would point the dish at the distant horizon where Pat would be driving through the beam of the telescope. We dutifully carried out this experiment, and I noted on the chart recorder when the pulses produced by his car came through. I could also hear the sounds of the ignition on the audio monitor attached to the receiver and there was no doubt it was his vehicle.

I do not recall whether Lovell managed to use these chart recordings to prevent the construction of the new subdivision. At the time, local politics did not concern me very much. However, I did feel a sense of pride in having made the first observations of the radio signals from a terrestrial automobile using the world's largest radio telescope.

## 9.5. Novae

The word nova means "new star." From time to time a new star seen with the naked eye does appear in our heavens. Such an event occurred in August 1975, when a nova appeared in the constellation of Cygnus the Swan. Nova Cygni became as bright as the Pole Star and then faded away after 8 days. For years radio astronomers monitored the dying radio signals from the Nova Cygni outburst.

The nova experience is quite unlike that of a supernova. The latter is the complete destruction of a star, while the former is the mere shrugging off of an outer layer of gas in a relatively minor convulsion, but one that would destroy our planet's atmosphere should the sun ever go nova, an unlikely event according to current theoretical knowledge of stellar evolution. Afterwards such a star may resume its normal existence or may repeat the process, in which case it is known as a recurrent nova. Radio emission from novae is produced in the ejected circumstellar envelope.

## 9.6. Other Superstars

Another marvelous phenomenon has been discovered in binary star systems. One star heats the particle wind blowing out from the other star. For example, the star Alpha Scorpii has two radio components associated with two stars in the binary. A point-like radio source is situated at the location of the smaller member and a small nebula is observed around the relatively more massive companion star. The nebula is produced by the ionization of the wind blowing from the smaller star as it moves past the larger one. The larger star does not produce a significant outflow of gas in its own stellar wind, but does generate a lot of ionizing, ultraviolet radiation. The smaller star produces a strong wind, but very little ultraviolet. Through teamwork they create a fascinating double radio source. This is an example of what is known as a symbiotic star.

# 10
# Radio Galaxies

## 10.1. On Finding Distances in Astronomy

When optical spectral lines produced by hot gas in distant galaxies is examined, a systematic shift in frequency with respect to laboratory measurements is found. The more distant a galaxy, the lower the frequency (the longer the wavelength) of the light received on earth. This is known as the redshift, with light waves being stretched to the redder part of the spectrum for the most distant objects. Those distances are independently measured through a variety of techniques involving variable stars and the observed properties of supernova explosions. Early in the 20th century Edwin Hubble defined this relationship and one of the principle goals of the Hubble Space Telescope was to pin down this redshift law, as it is called, more accurately. By knowing how redshift and distance are related, the redshift measured for any newly discovered object can be quickly converted into an accurate distance.

## 10.2. Chaos in Distant Galaxies

In 1918 a galaxy known as Messier 87 (33 million light-years distant in the constellation Virgo) was photographed and revealed a surprising jet of luminous matter emerging from its interior. In the late 1950s radio signals from M87 were discovered, one of the first examples of a mystery never before encountered—the radio galaxy. Why would an entire galaxy shine so brightly in the radio spectrum? Back then several other peculiar objects were identified with the strongest radio sources in the sky including two supernova remnants, the Crab Nebula (Figure 5.1) and Cas A (Figure 5.3), and a peculiar galaxy in the constellation Cygnus associated with the radio source named Cygnus A (Figure 1.4).

As radio astronomy technology improved, especially when a group of dishes are connected together to act as a single large radio telescope, many weaker radio sources were discovered and their positions measured. But the first positions were still too inaccurate to allow optical astronomers to identify an associated object, except in rare cases where something very obvious, and always peculiar, was visible at that location. Only in those cases did astronomers feel confident in relating the source of radio waves to the visible object.

## 10.3. The Largest "Things" in the Universe

In the early 1960s improved observations of distant radio sources revealed that the radio emission was often coming from two regions in space located on opposite sides of a faint, visible galaxy. The double radio source might be a minute of arc or so in extent with a much smaller (in angular size) galaxy located between the radio "blobs." Double radio sources were duly found to be common, but because of the poor resolution of those early radio telescopes little more could be said than that the radio source was a double. I recall endless discussions over lunch at Jodrell Bank in which we wondered why radio sources might be double. It soon became fashionable to invoke explosive events inside galaxies, which for some unknown reason ejected material in two directions. The central galaxy, if one could be seen at all, was often observed to have very active nucleus, inferred from the Doppler shifts of their light emission that implied chaotic motion.

An alternative explanation to account for the chaos in those distant radio sources was that galaxies were in collision, with each being torn asunder by their interaction.

Whichever idea one favored, it became apparent that in these radio galaxies immense amounts of radio, light, and even X-ray energy were being generated by dramatic events in the nucleus of what was usually the most massive member of a dense cluster of galaxies.

As a teenager I used to listen to the BBC on shortwave radio and one evening heard a talk about radio astronomy by Bernard Lovell, the Director at Jodrell Bank in England. I had never heard of radio astronomy, Jodrell Bank, or Lovell. During his talk he played a tape recording of what he claimed was the sound of colliding galaxies. I listened to the hiss of receiver noise, which gradually grew stronger and then weaker as the radio source Cyg A passed through the beam of the Jodrell Bank telescope. This stirred my imagination and about 8 years later I began working at Jodrell Bank as a graduate student.

## 10.4. Cygnus A

In 1953 the galaxy associated with Cygnus A, the second brightest radio source outside our solar system was identified. (The sun is the brightest radio source in the sky and the Cassiopeia A radio source the brightest outside the solar system.) The Very Large Array radiograph of Cyg A was shown in Figure 1.4 and it reveals magnificent detail. The radio lobes manifest as beautiful diaphanous filaments whose subtle patterns belie the amazing energies associated with this source. A faint yet stunning radio jet, less than a tenth of a percent as bright as the lobes, can be seen heading toward the northern lobe. The radio double is centered on a peculiar galaxy, which was originally believed to be galaxies in collision. In the 1950s two famous astronomers, Walter Baade and Rudolph Minkowski, argued about this, and bet a bottle of whisky or a thousand dollars, depending on whose version of the story you believe, on whether or not Cygnus A was a colliding galaxy. The issue was settled—against the colliding galaxy hypothesis—when it was realized that double radio sources were too common to be explained by intergalactic collisions. However, since then the explanation is again in question, because galactic cannibalism resulting from close encounters between galaxies in near collision within one another may be at work in many, if not all, radio galaxies. We must allow that Baade and Minkowski should both have won.

Several "hot spots" can be seen in the radio lobes in Figure 1.4. These are characteristic of many double radio sources and are often found at the end of the axes of the jets, where material crashes up against the boundary separating the radio lobe from intergalactic matter.

## 10.5.  The Radio Emitting Jets

From the depths of most radio galaxies, highly elongated and stable jets continually drive matter out into two enormous radio-emitting regions known as the radio source lobes. The new view of the physics and evolution of these immense radio sources suggest that these remarkable behemoths are not actually exploding, but continually spewing out incandescent matter in a steady stream that flows for millions, even hundreds of millions, of years. This hot matter is propelled outward from a black hole at the very heart of the active galaxy. As we shall see in the next chapter, it is the role of gigantic black holes that is key to understanding the radio galaxies.

Figure 10.1 shows the radiograph of a typical radio galaxy, known as 3C 31, which is associated with the visible galaxy NGC 383 located at the center of the image. (This nomenclature for radio sources is based on the Third Cambridge Catalog, painstakingly prepared by the Cambridge University radio astronomers after years of surveying the sky. Thus 3C31 is the 31st entry in their catalog.) To either side, two jets of radio emission blossom out into swaths of radiation that indicate that this radio galaxy is traveling through intergalactic matter that causes the ejected material to trail behind like the wake of a boat speeding through water.

Each radio source appears unique, yet underlying patterns emerge. They all show lobes of extended emission, far removed from the central galaxy/colliding galaxies. Most of them show jets, sometimes one-sided, sometimes two-sided, which are sometimes bent or swept back indicating motion through surrounding intergalactic gas. A radio galaxy radiates a million times more energy across the entire electromagnetic spectrum than does a normal galaxy, and its radio emission alone can outshine a spiral galaxy by 100,000 times.

Figure 10.2 is a radiograph of 3C 449, a radio source that is remarkable for its mirror symmetry (imagine placing a mirror vertically between the two sides of the

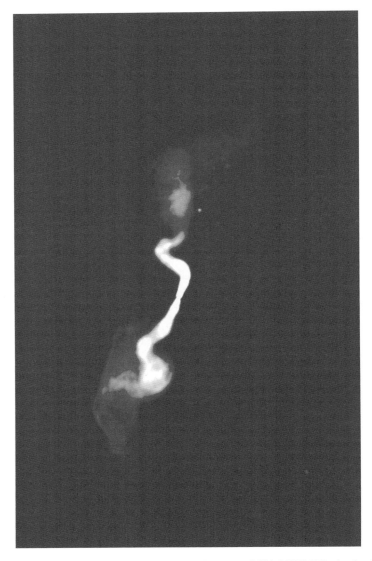

FIGURE 10.1. Spectacular twisting jets in the radio source 3C31 (NGC 383), the dominant galaxy in chain of galaxies, terminating in distorted, radio emitting plumes, which stretch to a distance of a million light-years from the center of the Galaxy. Investigators: Robert Laing, Alan Bridle, Richard Perley, Luigina Feretti, Gabriele Giovannini, and Paola Parma. (Image courtesy of NRAO/AUI.)

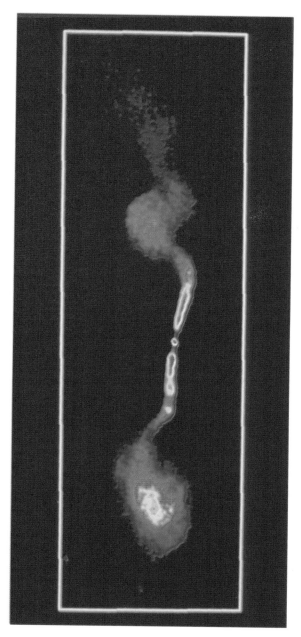

FIGURE 10.2. Very Large Array radiograph of 3C449, a radio galaxy located 230 million light-years away that shows two jets emerging from opposite sides of the nucleus of a large elliptical galaxy whose optical dimensions are about 1/10th of the inner radio features shown here. The radio jets expand to form diffuse radio lobes that trail behind the Galaxy as it moves through space. Investigator(s): R. A. Perley and A. G. Willis. (Image courtesy of NRAO/AUI.)

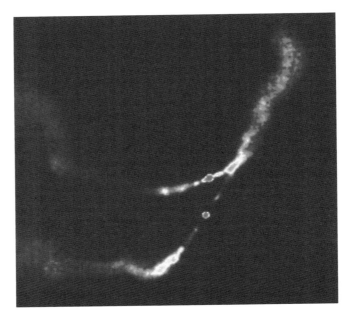

FIGURE 10.3. The 6-cm-wavelength radiograph of the source 3C75 with its interacting twin jet source associated with the central radio galaxy in a cluster of galaxies known as Abell 400. As shown in Figure, the jets are swept back as the galaxies move through space. On the right side of the image the two jets appear to interact and possibly are wrapped around one another, although they may also be located one behind the other. Investigator(s): F.N. Owen, C.P. O'Dea, M. Inoue, and J. Eilek. (Image courtesy of NRAO/AUI and F.N. Owen, C.P. O'Dea, M. Inoue, and J. Eilek.)

source). Swirls in the two extended lobes mimic each other yet they are several hundred thousand light-years in extent. For example, both jets show bends 300,000 light-years from the galaxy. Symmetry in the double-lobe structure appears to be common in radio galaxies.

3C 75 is the spectacular object shown in Figure 10.3. Four enormous radio jets blast out from two closely interacting/colliding galaxies. These lie at the center of a dense cluster of galaxies about 300 million light-years away. Each of two cores emits two jets that swirl and twist through space, to be swept back by the motion of the galaxies through surrounding intergalactic material. The jets appear to be intertwining in the northern parts although these are probably located one behind the other. The power associated with 3C 75 is 100 million times the energy output of our sun. These numbers are too staggering to comprehend fully.

The extent of the jets billowing out of 3C 75 is enormous—a million light-years across. The awesome amount of energy associated with those two jets, which have long since been expelled from the cores of their galaxies, implies that energy sources other than the original explosive energy of ejection must be operating in order to keep the material emitting radio signals for so long (many millions of years).

The presence of two bright cores in 3C 75, so closely spaced, is surely another example of galactic cannibalism at work. In this case two galaxies may have coalesced so that their two nuclei are now very close together, perhaps about to consume each other.

M87, known as Virgo A to radio astronomers, is one of the most spectacular of all the radio galaxies located relatively close to the Milky Way at a distance of 33 million light-years. Figure 10.4 shows a series of radio images at different resolutions. At the left the Hubble Space Telescope image shows an elliptical galaxy that exhibits a one-sided optical jet (not obvious here) that is itself a source of radio waves. The top left image shows that on the opposite side of the jet a radio lobe has inflated and that, in turn, is lost in the detail shown in the image at the right. The two stars in the top left image are shown in the right-hand image to indicate the changing scale. Also shown are lines that indicate the linear scale of the images

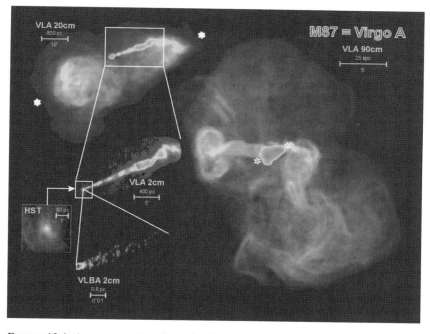

FIGURE 10.4. A sequence of radiographs showing ever more detail in the structure of the radio galaxy, M87 in Virgo (a.k.a. Virgo A). The enormous radio-emitting lobes are ultimately powered by a black hole at the galaxy's center, which lies deep within the bright, reddish region in this image. The structure in this larger image is approximately 200,000 light-years across. That is twice the diameter of the Milky Way galaxy. Investigators: F. Owen, J. Biretta, J. Eilek, and N. Kassim. (Image courtesy of NRAO/AUI and F. Owen, J. Biretta, and J. Eilek.)

FIGURE 10.5. Radiograph of the Fornax A radio source, centered on the giant elliptical galaxy, NGC1316 (center of the image), which is devouring its small northern neighbor. This Very Large Array image shows the radio emission to consist of two enormous radio lobes, each about 600,000 light-years across. The scale of such a structure is beyond human imagination, being a stunning six times the diameter of the Milky Way galaxy. Investigators: Ed Fomalont (NRAO), Ron Ekers (ATNF), Wil van Breugel, and Kate Ebneter (UC-Berkeley). Radio/Optical superposition by J. M. Uson. (Image courtesy of NRAO/AUI and J. M. Uson.)

in parsecs, bearing in mind that a parsec is equivalent to about 3.25 light-years. The whole of the Virgo A source produces an enormously complex swath of radio emission billowing out into space on a scale of 300,000 light-years, three times the diameter of the Milky Way.

Fornax A, a southern radio source associated with NGC 1316, is shown in Figure 10.5. In this image the radio emitting lobes have been overlain on an optical image of the central galaxy, again a giant elliptical, which is cannibalizing a smaller companion seen to its north. The enormous radio lobes are each about 600,000 light-years across. The matter blasting out from two jets travels a distance of 500,000 light-years before it is brought to a halt and splays out against the boundary of a lobe and illuminates the whole structure for astronomers across the universe to see. The collision is believed to have been going on for 100 million years. In such galaxies it is incomprehensible that any life would exist, given the intense gamma and X rays generated in the heat of collision.

The radio galaxy closest to us is known as Centaurus A (Cen A) located only 15 million light-years away. Because of its proximity it appears huge on the sky and no all-encompassing radiographs have yet been made that can be represented in one image. Its radio emission covers about 6° of angle and can be seen in the map of the radio sky shown in Figure 4.1, the elongated red feature about one third of the way in from the right, above the main band of Milky Way emission. This radio source is associated with an elliptical galaxy (which differs from a spiral galaxy in that it consists of densely packed stars with very little matter between them) known as NGC 5128 that shows a double, jet-like feature emerging from the heart of the galaxy. In the mid-nineteenth century words of Sir John Herschel, "cut asunder ... by a broad obscure band." This is an obscuring band of interstellar dust seldom, if ever, found in elliptical galaxies, but typical of spiral galaxies. Infrared observations of Cen A by the Spitzer Space Telescope showed that the dust is part of a spiral galaxy colliding with the elliptical, which is now consuming the interstellar matter of the spiral, an example of galactic cannibalism.

# 11
# Quasars

## 11.1. The Discovery of Quasars

While radio galaxies were often visible on optical photographs, back in the 1960s another type of radio source was found that appeared to have no optical counterpart. At the very best, astronomers might have noticed what appeared to be a star at the location of the radio source, but normal stars do not produce strong radio emissions. If those were stars they had to be a special class of objects and the name radio "star" was adopted, at least for a while, until one of the most amazing breakthroughs in the history of astronomy.

To set the scene, by 1963 the radio "star" 3C 48 had been located accurately enough in position to allow its photos to be taken. It looked like a star. In that year the moon happened to pass across an area of sky where another bright radio "star" 3C 273 was located. Because astronomers knew the moon's position very accurately they simply timed the disappearance and reappearance of the radio source as it was blocked (occulted) by the moon and converted the timing information into a position on the sky. The observations had to be repeated for another lunar occultation in order to resolve position ambiguity, because two points on the circular silhouette of the moon could produce the same time of disappearance. The second occultation would occur when the moon traveled along a slightly different path, and hence the timing information of two occultations could be combined to give an unambiguous position for the radio source. In this way the location of radio source 3C 273 was pinpointed by Cyril Hazard in Australia. More than that, the radio source was a double, just like so many radio galaxies had been found to be double sources. Here the two components were separated by 20 arcseconds.

This encouraged optical astronomers, in particular Maarten Schmidt at Palomar observatory, to photograph 3C 48 and 3C 273 as well as couple of other radio "stars" and to take their spectra. Around this time the class of objects came to be called quasi-stellar radio sources, quickly abbreviated to quasars. Determining what spectral lines were present in their spectra the mystery of the star-like radio sources would quickly be solved: were they stars, and if so what type of star?

The photo of 3C273 produced an incredible discovery. It did look like to be a star but it had a luminous jet protruding from it. The double radio source found by Hazard showed one located at the center of the star and the other at the tip of the visible jet.

At this point in the story the radio astronomical grapevine spread the news that the quasar spectral lines were utterly mysterious. These objects were not any known type of star, a discovery that set us all debating at length. What could possibly look like a star, produce tremendously strong radio waves, resemble double radio galaxies and yet produce so much energy? What made matters worse was that interferometer measurements of high accuracy showed the diameters of the quasars to be very small, which meant that they were surely not at the distances of galaxies because then their energy production would have to be even more prodigious with all of the energy coming from a tiny volume of space. That was also unacceptable to our imaginations.

It was up to Maarten Schmidt staring at the quasar spectra before him to try, after several months of getting nowhere, a bold new approach. The spectra did not look like any normal galaxy, but what if the spectra were subjected to a large redshift? Unimaginable as it seemed to at first that worked. The quasars 3C273 and 3C48 showed a large redshift, which meant they were not stars but possibly galaxies of some sort. If the redshift was an indicator of distance then 3C273 was about 1.5 billion light-years away. 3C48 was estimated to be at roughly the same distance, which corresponded to that of some of the most distant galaxies measured at the time.

By 1963 the mystery of the quasars had crashed about the astronomical community in full force. What sort of object could look like a star (even if one of them had a visible jet protruding from it) yet being farther away than most known galaxies? What could possibly be emitting so much energy as to shine more brightly than any other object in the universe? The energy problem was bad enough if these objects were stars in the Galaxy but for them to appear so bright while at huge distances seemed absurd.

## 11.2. Brightness Variations

The mystery became heightened when astronomers discovered that the light from quasar 3C279 changed in brightness over a year. Furthermore, a subsequent search of old photographs of 3C273 showed that it had undergone sudden changes in brightness over a period of 80 years. This incredible discovery was also completely unexpected. First, astronomers seldom observed time variability in objects other than stars within our Galaxy. Second, quasars are intensely luminous and are a problem in physics even if the energy is generated over volumes of space of galactic dimensions. This problem becomes far worse if the emitting region was very small. Variability of quasar brightness on a time scale of a year means that the size of the emitting region can only be about a light-year across, a limit set by the speed of light.

Soon after the discovery of optical variability in quasars, radio astronomers began to monitor them for radio variability, something that also seemed absurd at first, because most of the total energy from quasars was surely not originating in a volume as small as implied by the optical brightness variations. However, radio emission from quasars was found to vary in brightness from year to year, meaning that the luminous radio emitting regions were not merely very bright, but also very small. This posed an even worse problem for the theorists: how to explain huge quantities of energy being emitted in bursts from a tiny volume of space.

After finishing my graduate studies in 1965 and being taken on as a staff member at Jodrell Bank and the University of Manchester, I was asked to propose an experiment I would like to carry out. Optical variation of quasars had just been reported, and so I suggested I begin a program to determine if their radio brightness also varied. To put it politely, this suggestion was laughed off as absurd. I certainly could not justify doing this experiment on rational grounds, but then at the time quasars did not seem to be very rational creatures! Radio variability was subsequently measured by Bill Dent at the University of Michigan.

The full mystery of the quasar becomes more profound when all the information is considered together. They are very far away, and the luminous cores are very small. At the distances inferred by their redshifts, quasars emits the energy equivalent of a hundred billion stars like the sun, all the energy being generated in a volume of space not much larger than the solar system. Clearly impossible, or so it seemed, back then.

## 11.3. Parent Galaxies

For years controversy simmered as to whether quasars were isolated objects not associated with clusters of galaxies, or whether they were the cores of otherwise "normal" galaxies undergoing violent explosions. In the latter case surrounding galaxies may have been missed because the quasar so dramatically outshines it. A quasar may shine 100 times more brightly than a galaxy containing a hundred billion stars. (If a quasar phenomenon occurred at the center of our Milky Way, located 25,000 light-years away, the quasar would appear as bright as the moon.)

Painstaking research by many astronomers revealed that some quasars are indeed enormously luminous explosive nuclei of elliptical galaxies. These explosions are so bright that, because of the glare, we can barely see the surrounding galaxy. Only through use of very sophisticated photographic techniques and the largest optical telescopes have the parent galaxies been revealed in some of the closer quasars. Furthermore, because of the enormous distance to such objects, the surrounding clusters of galaxies are often invisible.

## 11.4. Quasars: The Modern View

Today hundreds of quasars have been identified and some of the most distant show redshifts equivalent to 95% of the speed of light, which places them 12–13 billion light-years away. We are seeing those objects as they were when the universe was barely a billion years old.

Observations with the Very Large Array show that quasars and radio galaxies look very much like one another with each showing jets feeding double radio lobes, as illustrated in the three images shown in Figures 11.1–11.4. Looking even deeper into the cores of the central objects, the quasars themselves, the Very Large Baseline Array shows blobs of radio emitting matter being blasted out at irregular intervals, the phenomenon that causes the optical and radio brightness to vary over time when the light or radio emission as a whole is observed.

The next chapter will show how the saga of the radio galaxies and quasars has been united into an overall, comprehensive view, one that none of us in the 1960s could have imagined.

FIGURE 11.1. Radiograph of quasar 3C204 showing the radio emission from relativistic streams of high energy particles fueled by events near a super-massive black hole at the center of the host galaxy (not shown in this image). The overall linear extent of the radio structure is 600,000 light-years. Investigators: Alan H. Bridle, David H. Hough, Colin J. Lonsdale, Jack O. Burns, and Robert A. Laing. (Image courtesy of NRAO/AUI.)

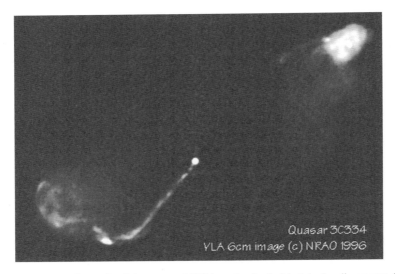

FIGURE 11.2. Radiograph of the quasar 3C334, a classic double-lobed radio source. Its linear extent is about 100,000 light-years. The southern jet is pointing toward us and appears much brighter than a faint counter-jet, due to the effect of having very high speed particles traveling out along the jets, as discussed in Chapter 12. Investigators: Alan H. Bridle, David H. Hough, Colin J. Lonsdale, Jack O. Burns, and Robert A. Laing. (Image courtesy of NRAO/AUI.)

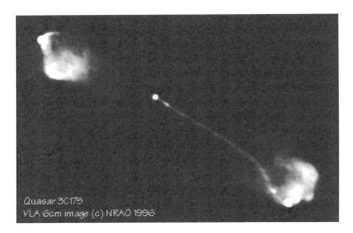

FIGURE 11.3. Radiograph of the quasar 3C 175, another classic double-lobed radio source. The overall linear size of this radio structure is nearly a million light-years or ten times the diameter of the Milky Way. The jet that is pointing toward us appears extremely bright because the particles emitting the radio radiation are moving toward us at close to the speed of light. Investigators: Alan Bridle, David Hough, Colin Lonsdale, Jack Burns, and Robert Laing. (Image courtesy of NRAO/AUI.)

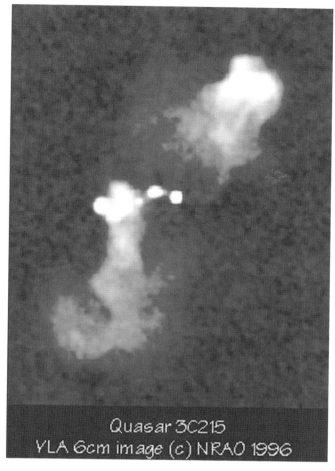

FIGURE 11.4. The radio emission from streams of high-energy particles traveling close to the speed of light generated by the quasar 3C215. The overall linear size of the radio structure is a staggering 700,000 light-years in extent and shows an unusual plume-like structure. The jet is extremely twisted and knotty and overall the structure is unusually distorted on all scales. Investigators: Alan H. Bridle, David H. Hough, Colin J. Lonsdale, Jack O. Burns, and Robert A. Laing. (Image courtesy of NRAO/AUI.)

# 12
# The Grand Unification: Active
# Galactic Nuclei

## 12.1. Cosmic Jets

The clue to understanding radio galaxies and quasars lies in their remarkable
radio emitting jets. It's all a matter of perspective. Long, narrow streams of highly
energetic gas squirt from the center of a galaxy, emitting radio waves as they go.
The radio jet in NGC 6251 is dramatically shown in radiograph form in Figure 12.1.
It is 1.2 million light-years in length, which makes it the straightest and longest
known object in the universe. The jet is a conduit along which energetic material
carries energy and magnetic fields from the nucleus of the galaxy to the outer radio
lobes. How are these jets formed and what holds them together? For the jets to
be so long and straight a good "memory" is required, something that allows the
flowing material to maintain a uniform direction for a very long time, a million
years or more, which would be the travel time of some of the jets if the matter
flowed at the speed of light. But how fast is that material flowing? Where does the
energy come from that enables them to illuminate radio lobes a million light-years
away? To rephrase this, what inflates the radio lobes?

Radio astronomers have been confronted with these problems ever since double
radio sources were first discovered; how can the radio sources emit so much energy,
and how can they do so for long periods of time? A fascinating new explanation has
recently been proposed. The jets obtain energy by transfusion in a most remarkable
manner.

Cosmic jets are the channels along which power is supplied from a galactic
nucleus to the extended radio source. The innermost radio jet in M87 (Figure 10.4)
is visible optically but most radio sources have jets that are invisible and nevertheless are radio emitters, sometimes showing associated X-ray emission. Jets are
believed to carry equal amounts of electrons and positively charged ions out into
the radio lobes, but only the electrons are observed, through the radio waves they
generate. Radio polarization observations reveal magnetic fields parallel to the jets
in the more powerful sources (when a jet is observed) and perpendicular to the jet
axis in the weaker sources, although even there they are parallel near the walls of
the jet. This magnetic field probably helps keep the jet under control and stable

FIGURE 12.1. Very Large Array image of the famous "blowtorch" jet in the galaxy NGC6251. The longest, straightest object discovered to date that shows patchy structure along its 1.2 million light-year length produced by shock waves as described in the text. (Image courtesy of NRAO/AUI.)

over long periods of time, otherwise the flow would become unstable—that is, lose its ordered structure—and quickly destroy itself.

When radio doubles were first studied it was suspected that something had to be flowing out from the galaxy in order to inflate the radio lobes. While no one specifically predicted that such narrow jets would be observed, and certainly not that they would be such beautifully organized radio emitting structures, their discovery has turned out to be yet another exciting topic in astronomy. It is the jets that funnel energy into the radio lobes.

Bulk kinetic energy (energy of motion of large masses of material in the jet) can be converted into particle acceleration in the jet through the action of turbulence (that is, chaotic motion) within the jet itself. Much of the energy may be generated within the jets as well as in the radio lobes and may not have originated near the central black hole. As the jets blast through space they draw energy from surrounding interstellar and intergalactic gases. To account for this, at least three mechanisms for adding energy to the relativistic particles have been proposed. Shocks, which are sudden discontinuities in the properties of the material involved, such as abrupt changes in density or motion, ahead of the jet and along its walls, create stronger magnetic fields, which in turn can accelerate particles. This process converts energy of bulk flow in the jet into relativistic energy of electrons; that is, the electrons are accelerated close to the speed of light. These particles, after radiating their energy, slow down to become nonrelativistic and hence no longer produce synchrotron emission.

Turbulence in the medium can also accelerate plasma. Electrons may be reenergized by collisions with protons in the plasma stream. Thus we have particles continually propelled back to close to the speed of light.

As the material streams outward, matter is dragged or sucked in from the medium surrounding the jet. This process, called entrainment, has interesting consequences. Whirlpools of matter, known as eddies, will be set up around the edge of a jet, just as in the wake of a ship. These eddies create shocks. The shocks can heat the gas to 10 million K and some of this gas then accumulates into dense pockets behind the shock. This entrained gas is predicted to cool to 10,000 K and show optical emission lines, which is observed in Cen A. By this time the gas will have moved as far as 30,000–300,000 light-years from the core. It is postulated that some of this gas will cool further, and stars will be formed, and this has been observed in Cen A. These stars will subsequently evolve, age, and die as they move out with the jet. Many will end their lives in explosive deaths known as supernovae and these explosions will, amazingly, become a significant source of new energy for the radio jets and lobes. Bear in mind that the material in the jets takes millions of years to reach the radio lobes, and therefore stars have plenty of time to be born and die during their journeys along these conduits into space. Entrainment, therefore, leads to a series of events, including star formation, which keeps the jet refueled and the radio source "shining."

## 12.2. Seyfert Galaxies

The radio galaxy and quasar story became clearer with the discovery of another class of galaxy that exhibit violence in their cores and which are relatively faint radio sources. The best known of these is a variety of spirals called Seyfert galaxies that show broad optical emission lines at their cores, the broadness indicating extremely chaotic, violent movement at the centers of those galaxies, motions which are not found in the more peaceful and normal spiral galaxies such as our own. Many Seyferts also show peculiar and distorted appearances on optical photographs. An example of a radio source centered on a Seyfert galaxy is shown in Figure 12.2. Their radio emission is usually confined to the core region and if a jet is present it is usually very short, less than a thousand light-years long because it cannot penetrate the surrounding gas and dust in the core of the spiral galaxy. Seyfert galaxies represent a class of objects that are neither a radio galaxy nor a quasar yet they exhibit chaotic motion in their cores.

## 12.3. The Energy Diet of a Jet

Radio source jets usually contain a series of brightly emitting "knots" as can be seen in Figure 12.1. This is expected because the jet does contain a magnetic field, which sometimes gets kinks or knots in it. Matter will pileup at a knot and cause a brightening, due to increases in the local magnetic field strength where more synchrotron emission is generated. This will cause cooling, just as anything that radiates away energy cools down, and so this region of the jet will contract, and possibly even collapse as it gets cooler. This collapse increases the field strength and

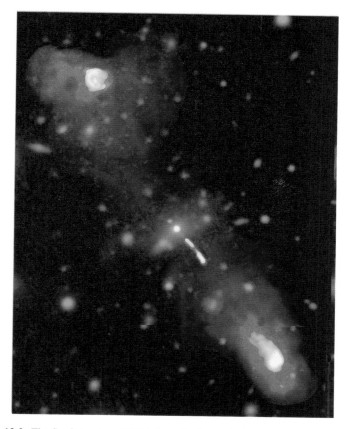

FIGURE 12.2. The Seyfert galaxy 3C219. Its radio image (red and yellow) is superimposed on an optical image. The overall size of the radio source is over 1 million light-years. The radio lobes are filled with networks of filaments and the core of the galaxy itself is a bright radio source. Investigators: David A. Clarke, Alan H. Bridle, Jack O. Burns, Richard A. Perley, and Michael L. Norman. (Image courtesy of NRAO/AUI.)

the emission. This cycle can continue indefinitely and creates an instability, which can play havoc with the smooth flow in the jet. The magnetic kinks may be expected to move with the flow. Should anything get in the way, such as a cold cloud in the surrounding galaxy (and some have been observed near the core of Centaurus-A), it will likely get swept up and accelerated until it is also part of the flow.

## 12.4. Faster than Light—Superluminal Motions

As we move in close to the core of a radio loud galaxy, we often find something even more incredible. Radio-emitting material blasts outward in bursts that have been tracked by very high-resolution interferometers with a resolving capability of a thousandth of a second of arc. The hot spots often appear to be moving faster than the speed of light! This occurs not just in the jet but very close to the core.

A stunning discovery about the bright radio cores is that they all show year-to-year movement at the smallest observable scales. Because the distance to a given quasar or radio galaxy is known from the redshift of its optical counterpart, it is possible to calculate how fast such blobs are moving. Depending on which source is being studied, they are found to travel between 3 and 20 times the speed of light! This conclusion, however, is at odds with one of the best known laws of physics—nothing can travel faster than light.

The discovery of this apparently superluminal (faster-than-light) motion in 1971 threw the astronomical community into a temporary tizzy. Relative order was restored by the realization that one can get the illusion of superluminal motion through a peculiar projection effect. If a double radio source is pointed nearly at us, then we obviously see the nearside material directed toward us and the farside material moving away. This has several consequences, some related to predictions made by relativity theory. The gas on the near side is blueshifted (because it is moving toward us) and that at the far side is redshifted (moving away). However, if the ejection velocity is close to the speed of light the emission on our side becomes greatly intensified due to relativistic effects, whereas the farside material would become so faint as to almost disappear. The apparent superluminal motion is a peculiar consequence of the fact that the material ejected toward us is traveling almost as fast as any light it emits toward us. A radio signal (traveling at the speed of light) cannot leave its source very far behind, and therefore two bursts of radio emission separated by a year could appear to us to be separated by a month, say, depending on the speed in the jet and on the angle between the jet and our line of vision. Therefore, when we see movement in the jets of the radio sources our initial estimate of the velocity of material could be completely wrong. This effect allows us to avoid the faster-than-light dilemma, but then another one pops up! If the large, straight jets are related to the small core jets, they too may be directed toward us, in which case they would be physically much longer than estimated. Therefore the jets may be far larger, and must be far longer-lived, than first suspected.

On the other hand, since the intensity of the emission from a core jet pointed nearly at us is highly dependent on subtle relativity effects, the energy we think it is emitting may be far less than originally estimated!

## 12.5. Active Galactic Nuclei

The manner in which all these inherently unbelievable and unlikely objects have been tied together in one grand unifying picture represents one of the great triumphs of astronomy in recent decades. The picture turns out to be surprisingly simple. It is mostly a matter of perspective; that is, it depends on how you look at it. Whether we see a quasar or a radio galaxy depends primarily on the direction in which the jets are oriented. If the jet is headed toward us we see a quasar. If the jet is oriented across the sky we see a radio galaxy. And whether or not the jet has to burrow its way out of a dusty or gaseous core of a galaxy determines whether one of the other members of the cosmic zoo rears its head in photographs taken from earth. That is where the Seyfert galaxies fit in.

The common denominator for all these magnificent structures is that there is an enormous amount of activity, chaos if you will, in the cores of those distant galaxies, and that chaos can usually be tied back to the goings on around a massive black hole at the centers of those galaxies. This gave rise to the umbrella description "active galactic nucleus" or AGN.

A quasar represents the extreme case of the radio jet emerging from an AGN that happens to be directed at us. It looks like a point source of radiation so bright as to dominate its parent galaxy, which explains why quasars originally appeared to be isolated objects in the heavens.

Unfortunately, there is yet another problem implied by the existence of core radio source jets that show superluminal motions. How come so many are pointed at us? Unless we occupy a favored position in the universe, those distant jets should be pointed in random directions as seen from our vantage point. After all, the jets do not know we are here. This created a problem until it was discovered that there are vast numbers of radio galaxies at large distances that also existed in the early universe. Only a fraction exhibit jets pointed in our direction. An observer in another part of the universe would see a different population of radio galaxies and quasars.

## 12.6.  Black Holes

So where does the flow of matter in the jets come from? What can eject two jets of material traveling nearly at the speed of light, and why would it continue to do this for millions of years? The answer comes from recognizing that the observed properties of the sources define the underlying nature of the hidden "engine" driving the radio source. Whatever it is, the engine shows two preferred directions, oppositely oriented in the sky. It is also very steady. Such a thing is a spinning object, acting like a gyroscope, which can keep spinning very steadily for very long periods of time unless acted on by some external force tending to pull it out of alignment.

The invisible object is actually very small, but enormously massive. The only type of astronomical object that can satisfy the demands of the observations is a gigantic black hole. A black hole containing 10 million solar masses would be 3 light-min across, or approximately the size of Venus' orbit about the sun. A black hole containing 5 billion solar masses, such as those believed to exist at the centers of some radio galaxies, may be 28 light-h across, or more than twice the size of the solar system.

A spinning black hole literally distorts the space around it, and any matter that comes relatively close will feel its tug, just as any particle feels the tug of gravitating objects in its neighborhood. For example, interstellar gas near the object will move inward and will first settle into an accretion disk, which spins around the black hole. Interactions between particles of gas will force them to settle into this disk and the same forces will cause the rapidly orbiting material to move gradually closer to the central hole. The gas will grow hotter as more and more energy is created due to collisions and interactions between particles in the swirling disk around the black hole. This gas will grow so hot, as much as a billion degree K, that it will actually expand and form a fat torus—a doughnut-shaped region—rapidly spinning around the black hole. This is illustrated in Figure 12.3. Inside this torus

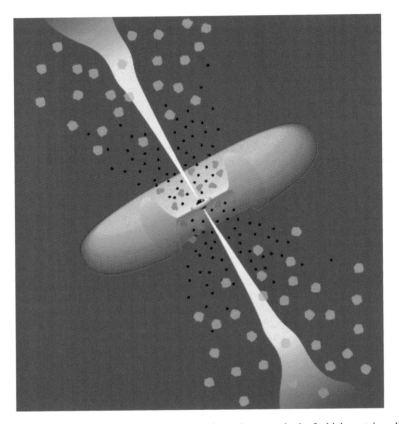

FIGURE 12.3. A schematic model for radio galaxies and quasars, both of which contain radio loud active galactic nuclei (AGNs), which explains most of the properties of these objects. A black hole at the core is surrounded by luminous material in an accretion disk shaped like a thick torus or doughnut. Narrow radio emitting jets of matter traveling close to the speed of light streak outward while small clouds of gas orbit the central region to produce optically observable broad emission lines, farther out and indicated by the light colored objects, other orbiting clouds produce narrow emission lines. As they fall in they will be swept into the accretion disk. (Image courtesy of C. Megan Urry.) This image originally appeared in the Publications of the Astronomical Society of the Pacific (Urry, C. M. and Padovani, P., 1995, PASP, 107, 803). Copyright 1995, Astronomical Society of the Pacific; reproduced with permission of the Editors.

will be magnetic fields that are literally tied to the black hole because some of the gas will have plunged into the hole and dragged the magnetic fields with it. Those magnetic fields at first remain connected to the gas outside and will rapidly wind up. Then, when they have become over-wound, they snap. As a result the fields will realign themselves, but since they are constantly being wound up they will snap again, and in this way energy from the rotating black hole is converted, through the magnetic field reconnection process, into the energy of particles where the field is so badly twisted and distorted. This process continues as long as the appetite of the black hole and the availability of gas allow it.

Around the poles of the black hole there is a critical funnel-shaped region of space in which matter finds it has two options. If it has insufficient energy it will plunge into the black hole and wave the universe goodbye. If, however, it has enough energy the particle may suddenly be free to escape from the funnel and blast out into space! This tends to happen in a series of outbursts in which blobs of matter are driven outward at near the speed of light.

If the shape of this funnel is narrow enough, provided the torus of gas around the black hole is thick enough, this matter will escape as if ejected from a nozzle (Figure 12.3). A similar beam of high-energy particles leaving the nucleus of a spiral galaxy would tend to collide with the surrounding interstellar gas, so abundant in spiral galaxies, and this gas would obstruct the flow. Hence black holes at the centers of spiral galaxies don't usually create jet-like radio sources. They are more likely to be observed as Seyfert galaxies. Only in relatively gas- and dust-free elliptical galaxies will the gas stream outward and be likely to escape unimpeded, as observed in classic radio galaxies and quasars.

The energy of this ejected material comes from the black hole itself. Based on an efficiency of 10% for the energy generation process, the most active galactic nuclei in distant radio galaxies must have processed the equivalent mass of 100 million suns through a region not much larger than the solar system. Since the sources are believed to have lifetimes of about 100 million years, it requires only one solar mass to be processed per year and only a fraction of that escapes up the funnel. That is enough to cause the AGN to glow. The black hole has to be about 100 million solar masses to do the job of creating a double radio source.

## 12.7. Precession

The jets of some radio sources show a wavy jet structure, almost as if the ejection occurred out of a hose, which was being swung around in a circle. It is believed that this is precisely what would be expected if the black hole is precessing, that is, if its axis is slowly wobbling about its average direction in space. A black hole may precess because of a gravitational pull from another object, perhaps another black hole, a galaxy, or a dead quasar passing close by. The flailing jets of 3C 75 (Figure 10.3) may be caused by the precession of a disk surrounding two interacting black holes. The precession is extremely slow and takes many millions of years to complete one wobble.

## 12.8. Galactic Cannibalism

Given that we have a black hole in an AGN doing all this wondrous stuff, we have to ask where the gas comes from that continually fuels the ejection from the black hole. That story again involves galactic cannibalism.

In the early universe, or in dense clusters of galaxies in which radio galaxies are found, galaxies are spaced very close together. They sometimes wander into each other's gravitational pull and become linked in an embrace that inevitably lures one galaxy into the other's cooking pot. We imagine one galaxy as a cannibal and a smaller galaxy, the missionary, tentatively but inexorably, approaching it. If the missionary gets too close it will get stripped of its gases. When the missionary falls into the cannibal's cauldron an AGN results. This is the cosmic feast that provides vast amounts of material that will be gathered into the accretion disk around the black hole. The fatal embrace between the two galaxies is long and slow, just what is needed to continually feed the black hole a steady diet of fresh gas. Jets are then blasted into space through the funnel around the black hole, allowing a reprieve for a small percentage of the captured particles.

It is also significant that the clusters containing the great numbers of galaxies (rich clusters) often have the most powerful radio sources near their centers. These are regions of overcrowding, just where cannibalism is most likely to occur. It is here that tidal encounters are more likely and these create frictional forces, which cause galaxies to spiral inward toward one another until they are close enough to gobble up one another. The most massive ones would form huge star densities at their cores, star clouds, which would provide an endless and systematic repast for the gourmet black hole.

Stars themselves may be involved occasionally. Old stars and exploding stars may provide an alternative source for material to feed to the black hole. The brightening of the cores, the optical variability, and the ejection of blobs of material along the jet in the form of superluminal motions may be related to sudden bursts of matter consumption on the part of the black hole, perhaps a sign that a star or two went down the wrong way.

---

I have urged readers to exercise their imaginations in order to comprehend some of the remarkable phenomena unveiled through radio astronomical observations. But can we really begin to grasp the enormity of what has been reported in these last few chapters. Galaxies colliding enormous black holes containing the equivalent mass of a hundred million suns, jets of material blasting out at nearly the speed of light to give us the impression, from our very large and very safe distance, of faster than light motion. Double radio sources where jets extend hundreds of thousands of light-years into space, in some cases in a straight path, in others swept back like wake behind a ship as the galaxy speeds through intergalactic space. Active galactic nuclei that tell of chaotic motion triggered as galaxies collide, and black holes at their cores drag matter hither and thither in swirling accretion disks heated to million of degrees to blast out

X rays and gamma rays that travel into the farthest reaches of the observable universe to alert distant astronomers to events that are virtually beyond imagination. We are very fortunate that our Galaxy harbors only a small black hole at its center and that there is no collision with another galaxy occurring right now, for otherwise life in the Milky Way would be very unlikely.

# 13
# Beyond the Quasars—Radio Cosmology

## 13.1. A Cosmic Perspective

When it comes to confronting the challenge that seems to fascinate us so profoundly, the origin of the universe, the imagination boggles for we have so little to hold onto. For the purposes of this chapter I will assume that the universe did have a beginning (the Big Bang), because to imagine that it did not is even more bizarre. The alternative, that it has lasted forever in some form, is something we cannot really comprehend. Just try! So we prefer to think of it as having had a moment of creation. And if it did, what experiment can we design in order to "see" back to the beginning of time?

We begin with the fact that astronomers know that the universe is expanding. This conviction is based on the observation of a universal redshift phenomenon. Distant galaxies, whose distances are derived in a variety of ways, are moving away from us and the galaxies farthest away are receding most rapidly. This expansion of the universe is quite literally that. It is not just that the galaxies are moving apart in space, but it is space that is stretching out between them. It is empty space that is expanding.

We can next ask when the expansion began, which is like asking when the universe came into existence. The answer is found by calculating how long it would take if one reversed the present expansion for everything to have been in one place. The answer, based on a great deal of new data obtained with the Hubble Space telescope, is that the universe is about 13,700,000,000 years old (13.7 billion). For comparison, the earth is about 4.7 billion years old, as inferred from dating of terrestrial and lunar rocks, and meteorites.

That is a long time compared to the 12,000 years that have passed since the last ice age held the planet in its grip and a few lucky human beings lived in caves, sheltering from the cold. Viewed from a cosmic perspective, 12,000 years is nothing—about one-millionth of the age of the universe. If the 13.7 billion years since the Big Bang were compressed into a single "cosmic year," a century of our time would be equivalent to one-fifth of a cosmic second, which means that the dawn of the era of modern science in 1600 or so happened about one cosmic second ago. This may give us pause to ask whether it is possible that we have learned so much in this instant of cosmic time that we may be close to the "final truths" about the nature of the universe.

It should come as no surprise that our brains, when pitted against this extraordinary universe, might take longer than a fraction of a cosmic second to come up with all the answers to our questions concerning how we came to exist on this planet, and how the universe came into being in the first place. Therefore, as we wander into the depths of space and time, do not be surprised if nature may yet have more to teach us as we struggle with the question of our origins that have for millennia not only inspired us to evolve as thinking beings, but which (unfortunately) have also caused us to band together and kill each other in the name of our cherished beliefs.

## 13.2. Radio Astronomy and Cosmology

While the picture outlined in previous chapters to account for quasars and radio galaxies makes a lot of sense, it is not at all clear that all aspects of cosmology make a lot of sense—not yet at least. If you happen to disagree with this statement turn it around and examine the implications. If you do believe that are close to a full and comprehensive understanding of the origin and evolution of the universe, what will there be left for astronomers to do for the next 100 or 1,000 years, other than fill in countless little details?

Radio astronomical observations can contribute to our knowledge of the universe on the largest scale, which implies that if the universe is expanding it might be possible to discover whether radio sources were different in the early universe as compared with our present epoch. This expectation produced high hopes in the 1960s, when quasars and radio galaxies were first discovered and radio cosmology research began with great enthusiasm.

When radio sources still appeared as single points of radio emission, it was assumed that they were intrinsically the same sorts of objects, and hence the distribution of sources over the sky could be studied. However, the radiographs and contour maps of distant radio sources shown in this book now illustrate the point that radio sources show tremendous variety. These objects do not represent a simple, uniform sample of objects. On the contrary, the distribution of radio sources in space does not appear to lend itself to any broad-brush interpretations, so they may not contribute to the hoped for cosmological study. In fact, there is no unambiguous evidence to suggest that any aspect of radio sources depends on their distance, that is, on the epoch in which the sources were born. The only thing we know is that quasars, in general, can be seen farther away than radio galaxies if only because the radiation beamed along the direction of a jet oriented our way happens to be enhanced by relativistic effects. Also, in the early universe all matter was gathered more closely together, and hence galaxies had a greater chance of colliding, so that active galactic nuclei might be more common back then.

## 13.3.  The Microwave Background

There is one type of radio observation that contributes decisive input to cosmological discussions. In 1963, Arno Penzias and Robert Wilson, at Bell Laboratories, which by coincidence was the birthplace of radio astronomy, revealed that the universe around us is bathed in a 3 K glow of microwaves. We are all stewing in this mellow glow, which represents a faint memory of conditions dating back to soon after to the creation of the universe. The discovery of the microwave background is an interesting case history in scientific progress, because when the radio signal was first detected it was suspected as being due to some spurious effects in the radio receiving system itself.

The story begins before World War II, when the physicist George Gamow, interested in explaining how the various elements came to be formed, hypothesized that the universe started as a fireball in which the elements were cooked up. Much later, Robert Dicke and his colleagues at Princeton University were planning observations to search for evidence to indicate whether we live in an oscillating universe that may have gone through a hot phase. Penzias and Wilson, meanwhile, were doing engineering work with a very sensitive radio antenna at Bell Laboratories.

Arno Penzias had been drawn to Bell Laboratories in 1961 with the promise that if he could get a horn antenna (a satellite communication antenna used in the Telstar project to send TV signals across the Atlantic) to operate at really high efficiency he could later use the antenna for radio astronomy. Robert Wilson joined him soon after, and together they were able to eliminate many sources of noise which lessened the efficiency of the antenna. When they made a list of all sources of noise, which included the sky, the ground, cables, and the radio receiver connected to the antenna, they were left with some residual noise, equivalent to that which would be produced by a thermal radiator with a temperature of about 3 K. In order to eliminate this they explored further and took the horn antenna apart. Two pigeons, contributing about half a degree of the mystery signal, were nesting inside. The pigeons were shipped out of town, but they soon returned—they were homing pigeons! After the birds were barred from entering the horn antenna again the astronomers returned to their measurements, because they still had nearly 3 K of mystery signal to contend with.

Penzias and Wilson believed that space was empty and shouldn't be radiating at 3 K. They did not know that Dicke was searching for evidence that the ancient universe might be hot, nor did they know that Gamow had wanted a hot universe to cook elements. Meanwhile, two Soviet scientists, A. G. Dovoshkevich and I. D. Novikov, suggested that if certain theories of element formation were correct, the early universe should be radiating a detectable radio signal. This work was unknown to any of the above, largely because of the poor communications between east and west during the Cold War. So there they were! Penzias and Wilson found a signal, but didn't know what it was, and Dicke and his team were preparing to search for such a signal and didn't know it had been found! At first they wanted to publish their discovery but suspected that it might be due to an error of measurement. For a while they were torn between not publishing the data or relegating their report to the relative obscurity of a technical, nonastronomical publication about antennas.

Meanwhile, another astronomer, James Peebles, independently predicted that a radio signal of 10 K should be expected from the Big Bang. As a result of conversations between astronomers and physicists in different lunchrooms, libraries, and offices, personal connections were made and the two Bell scientists realized what they might have discovered. The 3 K signal was probably real after all, and not due to pigeons or unexplained electrical problems in their antenna. It was a signal from the beginning of time. In 1965 Penzias and Wilson finally published their report in the Astrophysical Journal and used the title, "A Measurement of Excess Antenna Temperature at 4080 Megacycles per Second," displaying, to the end, extreme caution about the significance of their discovery. A theoretical explanation for the signal was given in the same journal by other scientists.

In 1978 Penzias and Wilson received the Nobel Prize in Physics for their important discovery. A twist to the story later uncovered in the documents at Bell Laboratories revealed that three other groups using the same antenna had also found evidence for this weak signal but none of them asked the basic question that drives scientific discovery: Why? Progress in understanding a mysterious new phenomenon requires that the question be asked in order to drive curiosity along the path that must be taken to find an answer.

The 3 K microwave background signal (now more accurately measured at 2.735 K.) appears to come from all directions in space. Light began its long journey to earth about 300,000 years after the Big Bang occurred, when all of the universe was at a temperature of 3,000 K. Because of the expansion of the universe, which causes the light from distant objects to appear redder than expected, the light from the Big Bang has been stretched to the point that it arrives here as a faint signal at radio wavelengths.

## 13.4. Beyond the Big Bang—Multiple Universes

Our book has concentrated on the invisible universe as revealed by radio astronomers, and it is beyond our scope to enter into a comprehensive description of the theoretical aspects of this science. However, at least one development in the understanding of cosmology has to be mentioned in order to explain the observations and relate them to what is known from laboratory measurements on fundamental particles. Our observable universe may be only one of a vast number of universes, all created in the Big Bang. The number of these possible universes is so vast that as far as our imagination is concerned it might as well be infinite. (The notion of infinity can really boggle the imagination!)

A new and more dramatic variation on the Big Bang theory has been suggested. The concept is known as the Inflationary Universe. The new cosmology can be related to events that occur in the laboratory, in particular to the spontaneous creation of particles in a vacuum. According to quantum theory, particles with very little energy can appear from nothing, exist for a brief moment, and then vanish. These virtual particles, also known as quantum fluctuations, are observed in laboratory experiments and may also appear spontaneously at any time. The greater the energy the shorter the lifetime of the particles and a system with exactly zero energy could, in principle, create particles from nothing that last forever! Hence we may have a way of creating a universe.

The essential point is that under certain conditions matter can be spontaneously created, and given the right initial conditions, this process can go on for a long period of time. The conditions for creating matter exist near a black hole, at what is known as its event horizon. This is the distance from the center of the black hole from within which no light can escape. The physical conditions at the event horizon are so extreme that virtual particles can spring into existence there. When they are created they usually disappear again as they cancel each other out. The physicist Stephen Hawking, however, showed that a black hole can evaporate, with virtual particles appearing near its boundary, some disappearing inside and others leaking into space. Such particles can be created not just at a black-hole edge, but at any event horizon. In the Big Bang these event horizons will have been present everywhere, and in vast numbers. They form "bubbles" of ordinary space–time. As soon as a bubble forms it expands at the speed of light while filling with dense matter by the Hawking process. As the bubble grows, its initial rapid growth gives way to a slower expansion and the creation of matter ceases. However, in the process a universe has been created, a universe that was originally a single bubble in a vast number of such bubbles, and which now appears to us as an expanding universe.

These theories predict that our universe may not be unique. If its creation happens once, a similar creation is possible many times. According to some cosmologists there may be countless universes, each disconnected from all the others. This implies that our observable universe is not all there is, but merely a sample of something far greater, something utterly unknown and unobservable.

Each universe is disconnected from all others because light cannot cross them fast enough to communicate about the existence of other universes before they have expanded out of sight. According to this view of the Big Bang, it appears that the universe within the observing range of our telescopes had a diameter of 1 cm at the end of the inflationary phase, before the expanding universe we now observe came into existence. Another way of putting this is that as you look farther and farther into space surrounding us in all directions, you will ultimately come to a place where the universe was only 1 cm across, but it is all around us! This is impossible to imagine unless we hand-wave about curved space–time. Yet that is the nature of cosmology: It is very difficult to imagine.

## 13.5. How Smooth is Space?

Radio astronomical observations have a bearing on our picture of what happened in the early universe. After the inflationary period each universe would have been very dense and very smooth. But the present day universe around us is not smooth at all. Great patches of matter, called galaxies, are contained in larger irregular volumes of space containing clusters of galaxies. Enormous voids exist between them. How did this patchiness come about? Without some gravitational irregularity in the early bubble (which became our universe) there would have been no seeds from which to grow galaxies. Such seeds should have existed as irregularities in the early universe, which should be detectable as slight nonuniformities in the microwave background.

It took several epoch-making experiments conducted with space-borne radio telescopes to home in on the smoothness of the microwave background radiation. The first, called COBE (Cosmic Background Explorer), set everyone back on their heels by finding that the universe was far smoother than expected, which made the problem of formation of galaxies out of the homogeneous material in the early universe nearly insurmountable. It was that experiment that caused one of the principal investigators to proclaim on national television with embarrassing (and absurd) hyperbole to having "seen the face of God."

The more recent and widely publicized mission of WMAP (Wilkinson Microwave Anisotropy Probe) has pushed the euphoria to new heights, as regards the extent to which the data have been interpreted. The essence of these experiments was that the WMAP science team made all-sky maps at various radio wavelengths (from space, in order to avoid problems with the atmosphere getting in the way) and then corrected the data for the emission produced by foreground sources such as the Milky Way itself. This requires a full understanding of all there is to know about radiation from the Milky Way, the nebulae in the Galaxy, and from distant galaxies. The result is shown in Figure 13.1. This map has been widely heralded as showing the structure of the universe when it was something like 300,000 years old or perhaps as much as 1 million years.

---

As a professional radio astronomer writing about my science it is difficult not to consider the implications of what I am writing about. This is particularly true in the case the WMAP observations of the microwave background. A careful look at Figure 13.1, which is on the same coordinate scheme used in Figures 5.1 and 6.1, reveals a string of point sources that precisely follows the galactic plane near the map center. Two of those points correspond to the location of well-known HII regions in the Milky Way. This suggests that the corrections to the raw data were not as effective as the WMAP team may have hoped. After noticing this, and also suspecting that other galactic features seem to be present in Figure 13.1, I have learned that the WMAP team allow that their corrections of the radiation in the general direction of the galactic plane leave much to be desired. They now focus their discussions on other parts of the sky. Call me a skeptic, but in that case how certain can they be that their final map is a true representation of what the early universe looked like 13 billion years ago? For example, it doesn't take much imagination to identify patterns in Figure 13.1 that remind us of the spurs of galactic radio emission seen in Figures 5.1 and 6.1. Also, to those well versed in the nature of galactic radio emission, the bright red patches to the left center of the map appear to be associated with regions of high polarization of galactic radio waves. Readers may wish to extend such comparisons for themselves. The only way to be more certain as to

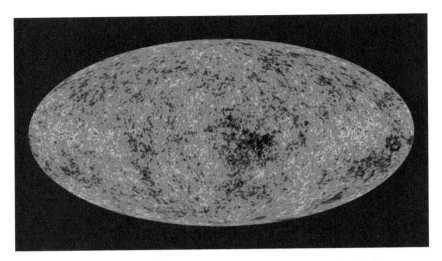

FIGURE 13.1. This all-sky map of the microwave background was obtained using a space-craft carrying a radio telescope as part of the WMAP. The patchiness is alleged to be due to temperature fluctuations that existed back when the Universe was less than a billion years old and which are now at a distance of 13+ billon light years from us. These irregularities may have acted as the seeds for Galaxy cluster formation. However, the reader may wish to compare this map with those shown in Figures 4.1 and 5.1 to see if you can recognize simi-larities. If this figure shows structure pertaining to the universe as it was 13 billion years ago there should be no similarity, because the other two maps represent radiation from matter in the Milky Way, on our doorstep. (Image courtesy of NASA/WMAP science team.)

the significance of the structures seen in Figure 13.1 is for an independent group of scientists to process the raw data using their own hypotheses as to what the corrections should be like. But that is an enormous task and it seems unlikely that any group will undertake such a venture, given that there is no significant reward to be had at the end. That, in turn, raises important issues that bear on how the scientific method works. Significant new discoveries are expected to be confirmed by independent studies. Might this no longer be possible for some of these enormous, dedicated space science programs?

## 13.6. Missing Mass (Dark Matter?)

Hydrogen makes up 80% of the universe's mass (the rest being mostly helium) and it was formed soon after the Big Bang. As the universe expanded and cooled, great volumes of this elemental gas formed enormous clouds, millions of light-years across. These systematically shuddered into many smaller clouds, each only a few hundred thousand light-years in diameter, later to become the galaxies.

Within galaxies, stars formed in great numbers and even smaller subunits broke free of the larger clouds and continued to collapse. Each collapsing cloud was pulled inward by gravity, which finally overcame the disruptive force of heat, that is, the internal energy of motion (known as kinetic energy). This process, a constant battle between gravity, which pulls clouds in upon themselves, and kinetic energy, which forces the cloud to expand, is witnessed everywhere in our universe. When gravity dominates, the cloud collapses to form a cluster of galaxies or a star, depending on the initial size of the cloud. Gravity will only overcome the disruptive influence of internal heat if the cloud cools by radiating away some of its energy as, for example, infrared radiation.

As the universe expanded and cooled, gravity became master of the destiny of gas clouds and determined the shape of things to come. Those shapes are what we now call galaxies, star clusters, stars, and planets. But there is a gigantic bug in the ointment. When the amount of observable mass in a cluster of galaxies is added together its gravitational pull is not large enough to hold the cluster together, given that the velocities of the individual galaxies have also been measured. The cluster should fly apart and it would not exist today, billion of years later. So what holds a cluster together? The most popular answer is that there is a form of dark, invisible matter that we cannot detect in other ways. Its presence can only be inferred from the pull it exerts on galaxies gathered together in clusters.

When it comes to the universe as a whole, astronomers would also like to know just how much matter it contains, because this will tell them whether the universe will go on expanding forever, or whether it contains enough matter to slow and later reverse the expansion. In the latter case the universe will collapse someday, leading to a catastrophic end to all that exists. Which leads to the bizarre new concept forced upon cosmologists because the most distant galaxies appear to be moving more rapidly than expected from the simple Big Bang model. That means they are accelerating, which requires that some hidden form of energy, known as dark energy, as the cause. Dark matter and dark energy? As Alice in Wonderland would say, "Curiouser and curiouser!"

---

At the time of writing this chapter I read an interesting alternative view of the "missing mass" problem that required no dark matter at all. The authors calculated how the members of a cluster of galaxies would move relative to one another if one applied Einstein's theory for gravity rather than the simplistic Newtonian version that astronomers had been using. This represented the first serious attempt to avoid the very uncomfortable view that the universe is filled with vastly more mass than we are capable of detecting through the radiations it might emit.

---

## 13.7. Gravitational Lenses

A tangible cosmologically interesting phenomenon concerns the gravitational bending of light. This was predicted by Einstein as part of his theory of relativity and was provided its first test when, in 1919, the British astronomer Sir Arthur Eddington measured the deflection of starlight passing close to the edge of the sun during an eclipse. The effect has also been checked by observing the change in position of distant radio sources observed in directions close to the sun's limb. The apparent shift in position is due to the bending of space by the sun's gravity.

A distant object such as a galaxy can bend the path of light or radio waves from an even more distant object. If the background object, such as a quasar, happens to line up with a foreground galaxy and the earth, the effect might be observed.

A gravitational lens concentrates the light (or radio waves) along a line rather than to a single point as glass lenses do for light. A star, galaxy, or quasar can form a number of images of a background object, with the images displaced from the object's true position, if it lies just to one side of the lensing galaxy. The image might also appear brighter than the original object. Depending on the relative orientation with respect to the earth, the gravitational lens can produce what is known as an "Einstein Ring," first proposed by Einstein in 1936.

Figure 13.2 shows the radio source, 4C 05.51, which exhibits a ring-like structure and two opposed compact sources. A massive foreground object, too faint to be seen optically, appears to be focusing the radio waves from a more distant quasar or radio galaxy.

An interesting and even tantalizing idea has emerged from the study of gravitational lenses. This phenomenon is believed by some to explain a peculiar aspect

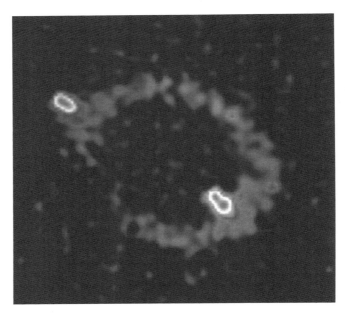

FIGURE 13.2. Radiograph of the unusual source, 4C 05.51, believed to be an Einstein ring produced by gravitational lensing of radio emission from a very distant background source by a massive foreground object. The two bright spots are further evidence that a lens is involved. Investigators: J. N. Hewitt and E. L. Turner. (Image courtesy of NRAO/AUI.)

of quasar locations on the sky. Several dozen years ago Halton Arp claimed that quasars, on average, congregated more closely to certain large galaxies than expected on the basis of chance. He and some of his colleagues suggested that the quasars were ejected from the relatively nearby galaxies, which meant that they were not at cosmological (very large) distances. That was a very unpopular suggestion at the time. However, if quasar light is amplified as it zooms past a galaxy this might cause such quasars to stand out in an all-sky survey. This would create a selection effect that misleads one to thinking the quasars are associated with the galaxies. (This may also be an unpopular suggestion!)

# 14
# On The Radio Astronomical Quest For Extraterrestrial Intelligence

## 14.1. "And Now for Something Completely Different"

Human beings have long been fascinated by the possibility of the existence of extraterrestrial intelligence (ET). As long ago as 50 BC, Lucretius wrote:

*Nothing in the universe is unique and alone and therefore in other regions there must be other earths inhabited by different tribes of men and breeds of beasts.*

Today questions about the existence of intelligent creatures on distant planets have taken on a new degree of respectability, thanks to the explosive growth in astronomical knowledge about stars and planets, and our recently developed technological capability to communicate, in principle anyway, with civilizations on planets orbiting other stars in our corner of the Milky Way. In recent years no book on astronomy appears to have failed to discuss the search for extraterrestrial intelligence (SETI), but in view of our complete lack of knowledge about the nature of ET, such discussions, despite attempts to present them in a scientific light, are primarily based on speculation and personal belief laced with vast amounts of hope. The SETI program can only be approached from the point of view of pure exploration, not as a scientifically justified experiment in the usual sense of the word.

It is to astronomers that people turn when the matter of Extra Terrestrial Intelligence (ETI) is raised and radio astronomers are at the forefront because communication between inhabited planets is likely to be with radio waves because they are easy to generate and can be beamed to travel a long way. There is a region in the radio spectrum, around 1,420 MHz (21-cm wavelength), where the radio emissions from the universe (microwave background, galactic radiation, radio sources) produce the lowest background noise and where the absorption produced by intervening matter, including the earth's atmosphere, is at a minimum. Thus the 21-cm band appears to be the natural choice for interstellar broadcasts. Furthermore, the existence of the 21-cm line of interstellar hydrogen gas would be known to other astronomically oriented societies, hydrogen being the most fundamental building block in the universe, and therefore radio astronomers have focused on this band as the likely the one for interstellar calls.

I was personally involved in SETI in 1971–1973 using both the late 300- and 140-ft radio telescopes of the National Radio Astronomy Observatory operating at the 1420 MHz frequency of the hydrogen line. Mine was the first experiment that had the sensitivity to actually detect a twin civilization if one happened to be on a planet orbiting any one of ten nearby stars (such as Barnard's star or Tau Ceti). No such signal was found. However, if the alien scientists had been experiencing a transmitter malfunction or had been on a lunch-break when I was ready to receive their message, taking into account the time delay for the radio waves to reach earth, the big moment would have been lost. Since then others have spent thousands of hours of telescope time, worldwide, in a continuing and fruitless search for ET.

At the root of SETI lies an important challenge. Our book has revealed that an enormous variety of radio waves pervade space, but for some reason no signal from ETI has yet been accidentally detected. Why is this? There have been a few false alarms, such as the reception of pulsar signals, which were first suspected of being from Little Green Men, and the discovery of the emission from interstellar OH masers, which had all the hallmarks of a signal from ET (narrow bandwidth and time variable). These phenomena were relatively quickly accounted for by clever theoretical interpretations, which should give us cause to wonder whether real signals from ET might, someday, similarly be accounted for in terms of cosmic exotica.

The interest in ETI is, however, made more relevant because of the discovery of many types of molecules in space (see Chapter 7). Over 100 species of organic molecules as well as the water molecule are pervasive in the Milky Way. Alien life, if it is out there, will be based on the same chemistry that makes us tick. The real mystery is whether any ETs have evolved a degree of technological sophistication that allows them to search for their neighbors in space, or whether they wish to make their presence known by transmitting radio signals. In this regard, our own technological civilization has only recently evolved to the point of being able to communicate over interstellar distances, hence it is timely to ask whether there are others like us out there.

## 14.2. The Harsh Realities of the SETI Equation—A Modern Heresy

An elegant way to express the likelihood that there may be other civilizations like ours in the Galaxy is the SETI equation, or Drake's equation. It is referred to by many writers who touch on the subject of the search for extraterrestrial civilizations but few, if any, have bothered to look closely at its implications. That is what we will do now.

The SETI equation gives N, the number of civilizations in the Milky Way with which can communicate now, and it can be written as

$$N = R^* f_p f_e f_l f_{co} f_i f_c L,$$

where $R^*$ is the rate of star formation in the Galaxy in number of stars per year; $f_p$ is the fraction that of those that have planets; $f_e$ is the fraction that, in turn,have planets that are ecologically suitable for life. On a fraction of those on which primitive life does emerge ($f_l$), evolution will cause complex organisms ($f_{co}$) arise and yet another fraction of those will give harbor an intelligent species ($f_i$), only some of which may become communicative ($f_c$) over interstellar distances. $L$ is the typical lifetime of a technological civilization, which means the lifetime over which the civilization has the interest in communicating with aliens orbiting distant planets, and has the ability to do so.

This simple equation summarizes a stunning amount of unknown information. It also has two terms that include time; $R^*$, the rate of star formation in stars per year, and $L$, the lifetime of a communicative civilization in years. When multiplied out the year terms cancel and we are left with a number of civilizations that now exist in the Milky Way galaxy with whom we could possibly communicate.

In dozens of astronomy classes I have enjoyed an exercise in imagination by inviting the students to guess at the value of the varies factors in the Drake equation. I stress that any guess is as good or bad as any other but that there are certain facts about the origin and evolution of stars, planets, earth, and life on earth, in particular those that affect survival, which have a direct bearing on the remarkable and no doubt unique course that evolution has taken on earth. We should use such knowledge to guide our guesses. No class has ever obtained an optimistic value for $N$, the number of civilizations in the Milky Way with which we could, in principle, communicate now, using present technology. Several students in a class at the University of Maryland were once so amazed at what they came up with that one or more of them contacted their representatives in Washington, DC, to point out that investing taxpayer dollars in the search was not justified. Months later an angry message from the leader of the SETI program was relayed to me through several intermediaries which in effect said that I should refrain from telling my students to contact their representatives with negative statements about the SETI program. In retrospect I was sorry I hadn't done so! Suffice it to say there was a lot of politics involved in extracting money from Congress and NASA to fund SETI. In due course NASA cut the SETI program which is now run by an independent organization using funds raised privately. I have also learned that they no longer use the Drake equation to rationalize their program, and the reason will become apparent next.

Let's see what is involved in estimating the number $N$. The stellar birthrate is given approximately by the total number of stars in the Galaxy, $(2 \times 10^{11})$, divided by the age of the Galaxy ($2 \times 10^{10}$) (in years), so that $R^*$ is about 10 per year. For the purposes of our discussion it doesn't matter if you prefer 20 or 30 for this number, as it will become apparent since we are dealing with enormous uncertainties.

In recent years astronomers have been finding planets orbiting other stars. That suggests that the fraction of stars that have planets, $f_p$, could be set equal to 1, true if all stars have planets. If we are pessimistic, set this equal to 0.1 (one in ten). For each of the factors I will suggest an optimistic value and a pessimistic one.

What fraction of those planetary systems will contain a planet that is ecologically suitable for primitive life to emerge? From the data gathered so far on 144 other planetary systems and the 168 planets orbiting those stars, they all have highly elliptical orbits or they orbit very close to their parent star. Also, most of the detected planets are larger than Jupiter. To date none of them fits the term "habitable" in light of the extreme variations in their obits. At the very least we would expect inhabited planets to have a reasonable equable climate, which depends on the planet having a closely circular orbit about its parent star. This situation may change in the next decade but in the mean time, using available information, we may be optimistic in setting set $f_e$ as one in ten, or equal to 0.1. Pessimistically we might go for one in every hundred stars, or $f_e = 0.01$.

What about the fraction of those planets that are ecologically suitable for life on which simple life forms actually emerge, $f_l$? This is likely to be very high, given what we know about the extreme conditions on our planet where living organisms have been found to not only survive but to flourish (known as extremophiles, such as those that live inside rock, in hot springs, or at great depths in the ocean where light does not penetrate). So let's set $f_l = 1$, which means life is certain to emerge if it is given half a chance. I'd call this both an optimistic and a pessimistic estimate.

Now consider a brief excursion into the most widely held misunderstanding of this whole issue. Most people raised on a diet of science fiction immediately and unconsciously equate the term "extraterrestrial life" with "humanoid life." However, for most of the history of life on earth the species that did exist were very, very primitive. Only in the last 500 million years out of the 4.5 billion years since the planet formed have complex organisms existed here at all. As regards a humanoid species, we've been here a few million years at most. This means that the answer to a question concerning the existence of primitive extraterrestrial life will be utterly different from one related to the existence of extraterrestrial, technologically sophisticated species.

Returning to the equation, the next term is the fraction of those planets on which life does emerge where the organisms reach a level of complexity such as the one that was seen on our planet 500 million years ago in the so-called Cambrian explosion. Before that life on earth could best be described as primitive and that epoch saw life suddenly become very diverse with larger and more complex organisms creating new branches on the tree of life. Back then earth was already 90% of its present age and there is no reason to assume that this jump to complexity was inevitable. What happened to create this enormous leap forward? The transition must have involved a significant change in the environment, such as the one that might be brought about by continents breaking apart and drifting over the face of the earth. But what triggered that? Was it a huge asteroid impact perhaps, or mega-earthquakes? We don't know, but what we do know is that the emergence of complex organisms was not a foregone conclusion, certainly not on a planet where the environment does not undergo change.

By definition, evolution involves change, and on a planet in a stable orbit about its parent star with an unchanging surface and a stable climate, the emergence of new species, and in particular the development of complex organisms, is highly unlikely. To phrase this differently, we are only here, now, because of the precise, turbulent history of our planet and because of fortunate circumstances that enabled life to undergo enormous and favorable transitions at various epochs in the past, at least as seen from our point of view. (Dinosaurs might not agree!)

So many factors play a role here. Consider the earth's magnetic field: It played a role after the Cambrian explosion when creatures moved onto land, where they would be exposed to ultraviolet radiation from the sun. Fortunately the earth's magnetic field acts as a shield that protects us from the solar wind of particles that otherwise smash into the atmosphere and destroy the ozone layer in short order. The ozone layer, in turn, absorbs solar UV and hence land-roaming creatures can survive beneath its shielding blanket. Without a magnetic field as a shield life suffers extinction events and these have been noted in the fossil record when the earth's field reversed, which it has done on a regular basis every few hundred thousands years. We are overdue for another.

Then there is the business of the earth's moon, essential for life as we know it since it seems highly likely that the impact with a Mars-sized object that ejected the material to form the moon may also have caused the earth to tilt on its axis by a very fortuitous amount that allows for an equable climate over most of its surface. This is not the forum in which to endlessly discuss all the factors that play a role in driving evolution, so let's just hazard a guess as to the value of $f_{co}$. Let's offer two values, an optimistic 0.01 (one in a hundred) as opposed to a pessimistic 0.001 (one in a thousand) for the fraction of planets on which life emerged and where the evolution of complex organisms took place.

Of those planets on which complex organisms do emerge, what fraction gives rise to intelligent species ($f_i$)? Let's ignore for a moment that we have several intelligent species around. In fact, members of all species are as intelligent as they need to be in order to be a member of their species. Otherwise they wouldn't survive for very long. A mongoose needs to know it is a mongoose or it will try to partner with a reindeer, which is bad for the survival of its species! Consider for a moment our civilization. Countless complex phenomena have been at play that set the scene for the emergence of *Homo sapiens*. Ice ages came and went and we are a technological society because the most recent one ended about 11,000 year ago and global warming since then has kept us living quite comfortably. Looking further back in time, for conditions to be suitable for mammals to rise to prominence required that competitors for living space, such as the dinosaurs, had to be removed. A fortuitous collision with an asteroid 65 million years ago appears to have done that trick. Had it arrived an hour earlier or late the dinosaurs might still be roaming our world. After all, they managed to survive for 200 million years before they were wiped out. We've also been lucky that a similar impact event hasn't happened since then! So let's ignore all the difficulties and set $f_i = 0.01$ or $0.0001$ depending on our mood at the time. Try this yourself. Your guess is as good as anyone's!

Of those planets on which intelligence emerges, what fraction will witness the development of a species that learns to communicate over interstellar distances? Here we at east know something about what allowed our civilization to evolve to the point of having radio telescopes to search for ET. Most importantly, our technological society is built on cheap energy. We have been blessed with vast reserves of oil and a highly scientific culture that allowed us to understand and harness electricity and magnetism. Is it likely that all planets inhabited by intelligent species have something like vast supplies of oil? As regards understanding electricity and magnetism, we may have been fortunate because nature provided us with amber (used to generate static electricity) and lodestone (a natural magnet) that stimulated human curiosity 3,000 years ago. It was the quest for understanding of electricity and magnetism, and how they were related to one another, that lead to the beginnings of the scientific revolution, as I have argued in "Hidden Attraction: The History and Mystery of Magnetism." Imagine what life would be like on a planet that did not offer its inhabitants amber and lodestone to spark their curiosity. Would those civilizations even develop a technological society? If you want to believe they would, what would allow such a society to become aware of electricity and magnetism at all? Imagining a world that alien is not easy. So let us set $f_c$ as $0.01$ or $0.0001$ for the two ends of our estimates, ignoring a great deal more about other factors that played themselves out to assure our emergence into the 21st century capable of interplanetary travel and interstellar communication.

Next we try to guess at the value of $L$, the average lifetime of a technologically sophisticated species. This is defined as the period of time during which a species has the will and the ability to communicate over interstellar distances, and wants others to know they are there. We have lasted about 100 years so far, but will we make it to 1,000? Not necessarily. Many dire events could wipe out our species, not least our own stupidity, such as a nuclear holocaust. Or perhaps disease or an asteroid impact or super-volcanoes could do the trick. Let's be optimistic and set $L = 10,000$ years, bearing in mind that 10,000 years ago we were still living in the caves and we cannot begin to image what our species will be like that far in the future. Pessimistically we could set $L = 1,000$ years before a major epidemic wipes us out to the point where we have to start all over again.

Now multiply out the terms in the equation and we come up with two estimates for the number of civilizations we can possibly talk to now, somewhere in the Galaxy. The optimistic value leads to $N = 0.01$ $(10^{-2})$. That implies you would need to search 100 galaxies before finding another civilization, which means we are alone in the Milky Way!

The pessimistic values we guessed at produce $N = 10^{-11}$. That means there is no other technological civilization in a universe containing even a hundred billion ($10^{11}$) galaxies! That would imply that we are very, very alone in the entire universe.

Feel free to play with your own numbers. Those people involved with religious fervor in the SETI program no longer use Drake's equation because the harsh reality is that it is impossible to be optimistic about finding anyone to talk to. Carl Sagan used to argue that $L = 10$ million years, or one thousand times as long as our optimistic number above. That would bring $N$ up 10 technological civilizations in the Milky Way with which we could, in principle, communicate. But if evolution continues there is no way to imagine what life on earth will be like 10 million years from now other than to say that *Homo sapiens* as we now know it will not exist unless evolutionary change ceases. But if evolution stops, can a species retain its technological prowess? And so the argument goes on and on, ad infinitum.

A final note, if you prefer to be hugely optimistic and assume that a typical lifetime of a technological society is 10 million years, it is possible to calculate how far those 10 twin civilizations are apart from one another, on average. It is about 25,000 light-years!

The estimates above are loosely based on what a typical class of students estimated for the factors after they had taken a semester or two of astronomy. The most recent class came up with $N = -10^{-8}$, which implies that the nearest twin civilization resides somewhere in the nearest 100 million galaxies! The optimistic estimates over the years have been as high as 1,000 civilizations in the Galaxy but pessimistic (realistic?) estimates always place us in a unique position in the entire universe. This invariably comes as a shock to the students. What if that were true? Wouldn't that be interesting?

While working on this chapter I read in an otherwise respectable book involving the history of radio astronomy where some otherwise respected scientists too leave of their senses and seriously proposed that we could expect the lifetime of a typical civilization to be of the order of the age of the Galaxy, or about 10 billion years. By setting the lifetime $L = 10$ billon years, the Drake equation implies that the Galaxy would be populated with twin civilizations! That is the only way of avoiding very low values found above for $N$. However it seems to me, a born-again skeptic about the value of the SETI program, that to suggest with a straight face that the typical lifetime of a technological civilization will be billions of years is so bizarre as to be absurd no matter how one looks at it. I'd bet anything that we will not be around as a recognizable hominid species even a million years from now!

I think it would behoove us to adopt as a working hypothesis that we are alone in the Galaxy. I have no doubt that planets all over the Milky Way are swarming with alien life, which, by definition, is alien. The vast majority of ecologically suitable planets will, at best, harbor organisms that we would call "extremophiles." Instead of fostering a new religion based on a belief that we will soon contact ET, we might more fruitfully attempt to handle the notion of being a unique species in the Milky Way. Perhaps then we would take seriously the issue of our continued survival in the face of human frailty and ignorance. We also need to take more seriously the likely consequence of some of options that nature has in store for us, because nature does not need our help to usher us off this stage of life we call Earth.

# 15
# Radio Telescopes: The Future

It was a beautiful sunny day in a Maryland suburb of Washington, DC. I was working on my home computer in preparation for an observing run on the 300-ft diameter radio telescope at Green Bank, West Virginia. The phone rang. It was my wife, calling from the Goddard Space Flight Center where she was working at the time.

"Have you heard what happened to the 300-ft?" she asked.

Surprised by this unexpected question I tried to imagine what would be the most inappropriate answer and responded, "Has it fallen down?"

"So you have heard."

"Heard what?"

"It collapsed last night."

"I don't believe you," I exclaimed.

But it was true. During the evening shift the telescope operator housed in the control room beside the telescope heard a very strange, sighing sound, almost ghostlike, and very loud. In a panic he rushed out of the building and headed to the control room of the 140-ft telescope a good half-mile away. After being consoled by the operator on duty at the 140-ft, he drove back to the 300-ft and was utterly horrified to see it no longer looming above the control room against the dark hillside beyond. Instead, there it lay, a pile of twisted metal where once this giant had stood, gathering faint radio signals from the depths of space.

Months later the likely cause was diagnosed. Metal fatigue, combined with a patchwork of fixes of cracks over a number of years to strengthen the girders holding the telescope together, had caused stresses and strains to be transmitted within the structure to the point where a key element gave way. One thing led to another and the massive structure sagged to the ground with a sign and a groan, frightening the only witness nearly out of his wits.

And thus began a new saga in the history of the National Radio Astronomy Observatory (NRAO) in Green Bank as Senator Robert Byrd, Chair of the US Senate's Appropriations Committee, made it clear that something had to be erected to replace the ruin so as to assure the continued operation of the observatory in what was otherwise a struggling community in West Virginia.

And thus it came to pass that the Robert C. Byrd Green Bank Radio Telescope (the GBT), a 100-m diameter (330 ft) beauty, arose in its place at a cost of some $5 million, a radio telescope now used to probe the distribution of complex molecules between the stars and the ticking of mysterious distant clocks known as pulsars.

## 15.1. Bigger and Better

The quest to "see" the radio universe more clearly has led to the construction of ever-larger radio telescopes in order to improve resolution. Following the collapse of the old 300-ft diameter radio telescope of the NRAO (Figure 15.1) the Robert C. Byrd Green Bank Telescope was constructed to take its place (Figure 15.2). In

FIGURE 15.1. At 9:43 pm EST on Tuesday the 15th of November 1988, the 300-ft telescope in Green Bank collapsed. The collapse was due to the sudden failure of a key structural element—a large gusset plate in the box girder assembly that formed the main support for the antenna. This is a photograph of the 300-ft telescope taken on November 16, 1988 after the collapse. The loss of the 300-ft telescope resulted in the Green Bank Telescope project, Figure 15.2. (Photo Richard Porcas, courtesy of the NRAO/AUI.)

FIGURE 15.2. The Robert C. Byrd Green Bank Telescope of the NRAO in Green Bank WV. This beautiful 330-ft diameter (100-m) dish replaced the old 300-ft that collapsed under its own weight in 1988. The unipod design is meant to reduce unwanted reflections of radio signals from steel beams in the support structure holding up a secondary mirror that then reflects the radio waves to a focus behind the main dish. (Author's photo.)

addition, new breakthroughs in electronics have allowed more sensitive receivers to be built. A flock of new telescopes have recently been designed to operate in previously unexplored parts of the radio spectrum.

The larger the physical (collecting) area of a dish-shaped radio telescope, the greater its sensitivity and the fainter the radio signals it can detect. The 1,000-ft Arecibo dish in Puerto Rico, completed in 1963, covers 20 acres, larger than the combined area of all the other radio telescopes in the world. It is built in a natural depression in the hills south of Arecibo and required relatively little work to scoop out the terrain so that its enormous reflecting surface could be snugly suspended in the preshaped space.

The smoothness of the surface of any telescope mirror, whether it be an optical or radio telescope, determines how good a reflector it is. The better the reflector, the more of the gathered energy is accurately brought to a focus and the less goes to waste by bouncing off in random directions. The surface irregularities have to be smaller than one eighth of the observing wavelength in order for the mirror to perform well. A radio dish may look rough to the eye but to a radio wave it appears as a shiny mirror. After being resurfaced in the mid-1970s, the Arecibo dish operated down to 6-cm wavelength, which means that over its entire surface the irregularities are smaller than 1 cm. This was achieved by accurately adjusting each one of 38,778 panels to an accuracy of a few millimeters.

## 15.2. Low-Noise Receivers

The radio amplifiers, or receivers, connected to the antenna at the focus of the dish-shaped reflector have to be of extraordinary quality to match the effectiveness of the radio reflector. Internal noise generated in the radio amplifiers, and in the cables or other hardware that connect the antenna to the receiver, competes directly with the faint radio whispers from space. If the internal noise levels are too high, faint radio sources will not be detected. In the early days of radio astronomy vacuum tube receivers had internal noise temperatures equivalent of thousands of degrees. Then, in 1961, a device called a parametric amplifier became available. Its noise temperature was first 350 Kand then later development reduced it to around 200 KThis meant that the sensitivity of radio telescopes suddenly improved tenfold, because the sensitivity depends directly on the noise temperature of the receiver. Modern radio telescopes use extremely low noise, solid-state receivers to produce overall noise temperatures of 20 KRadio receiver technology is unlikely to go lower because of limitations that have little to do with engineering. The 2.7 Kmicrowave background sets the absolute limit, and atmospheric radiation contributes another 6 Kor telescopes on earth.

## 15.3. SMA—The Submillimeter Array

The Submillimeter Array (SMA) is producing valuable details about the inner workings of distant clouds of molecules. It operates in the frequency range from 180 to 900 GHz (corresponding to very short wavelengths from 1.7 to 0.7 mm) where it exploits a narrow radio window in the earth's atmosphere that is partially transparent to waves at these frequencies. At other frequencies the atmosphere blocks out submillimeter waves. The array's efficacy is increased by its location on the Mauna Ka in Hawaii at an altitude of 4,080 m (13,386 ft). The SMA is a joint venture of the Smithsonian Astrophysical Observatory and the Academia Sinica Institute of Astronomy and Astrophysics in Taiwan. The Taiwanese group funded two of the dishes in the array, in return for 15% of the observing time, so that the SMA now has eight antennas, each 6 m in diameter (Figure 15.3) that can be moved to any of 24 locations spread over an area 508 m across. The maximum resolution is in the range of 0.1–0.5 arcseconds, depending on the frequency chosen. The SMA is fully productive but a summary of its key discoveries remains to be written.

## 15.4. Planned Arrays

Several new array telescopes making use of aperture synthesis principles are in the planning or in early construction phases.

FIGURE 15.3. Telescopes making up the SMA standing at attention on the Mauna Ka mountain in Hawaii. (Image courtesy of NRAO/AUI.)

## 15.4.1. ALMA—The Atacama Large Millimeter/Submillimeter Array

In the quest to better understand the nature of molecular clouds and what drives the chemistry of the universe, an array of 64 dishes, each 12 m in diameter capable of operating from 30 to 1,950 MHz (wavelength range from 9.6 to 0.3 mm) is to be constructed at the heady altitude of 5,100 m (16,500 ft) on the Chajnantor plain of the Chilean Andes in the District of San Pedro de Atacama. Figure 15.4 shows an artist's impression of what it will look like. At that altitude there is little water vapor in the atmosphere overhead, which would otherwise absorb millimeter radio waves, and there is also much less oxygen to breathe. The telescopes will be spread over 15 km of the plains. ALMA has its own science center far away at a low altitude, at the NRAO headquarters in Charlottesville, VA. Very few technical people will be on site, given that the array will be remotely controlled.

ALMA will cost something like $60 million by the time it comes into full operation in 2012. Already 7 years of funding by the US National Science Foundation has allowed the science center to be built, people to be hired, and telescopes and receivers to be designed and constructed. Any project of this scope is only possible with international cooperation and here we find the NRAO and the US National Science Foundation NSF as well as the National Research Council of

FIGURE 15.4. Artist's conception of the antennas planned for the ALMA. (Image courtesy of NRAO/AUI and ESO.)

Canada, working with the European Southern Observatory (ESO), Spain, and the National Institutes of Natural Science in Japan, and, of course Chile, which gets 10% of the observing time. The Japanese collaborators are contributing 25% of the cost ($80 million). In July 2005 a contract for the first 32 dishes was signed.

One of the primary research projects on ALMA will be to study molecule formation in galaxies that existed when the universe was still young. It will also look into the heart of radio galaxies and quasars in the era when galaxies were just being formed for the first time. Much, much closer to home, it should be able to image comets and asteroids with amazing clarity and study objects in an orbit about the sun beyond the orbit of Neptune and Pluto, the so-called Kiper Belt objects. It should even be able to detect radio emission form the surface of distant "normal" stars and observe planetary formation processes in detail.

Initial science measurements should begin in 2007 with the full array expected to be operational in 2012.

A prototype of the radio dishes that will make up ALMA has been brought into operation (Figure 15.5). Kown as APEXthe Atacama Pathfinder Experiment, it began operation in July of 2005 and is operated independently of ALMA by several European scientific organizations including the German Max-Planck-Society, the Swedish Research Council and the European Southern Observatory. It uses a modified ALMA antenna and is located at the ALMA site. In order to work effectively and efficiently at the submillimeter wavelength range these 12-m dishes have to be smooth to better than 17 thousandths of a millimeter, or, in the words of

The APEX Telescope at Chajnantor

© ESO

FIGURE 15.5. APEXelescope, a prototype reflector for ALMA, now in operation at the ALMA site in Chile. Clearly the desert terrain at an altitude of over 15,000 ft does not look inviting as a place where visiting radio astronomers can relax between observing runs. The telescope is therefore remotely controlled as will be the case for the larger ALMA at the same site. (Image credit: ESO.)

an APEXpress release, less than one fifth of the thickness of a human hair. That gives the dish its shiny appearance.

## 15.4.2. LOFAR—Low Frequency Array

This radio astronomical tool represents a classical case of thinking outside the box, or in this case outside the dish. Instead of moving cumbersome masses of metal to point at various directions in the sky, a practical alternative at low frequencies is to use simple antennas that individually have no resolving power. All their signals will be collected in a central computer and by adding various time delays to the electrical paths from elements of this array it is possible to process the data so as to simulate a beam in the sky whose width will be determined by the extent of the array. The beam can also be pointed electronically. It is a very clever idea made even more innovative by having individual antennas consisting of little more than four supports in the shape of a pyramid planted in the ground holding up simple dipole antennas that make use of the radio reflecting properties of the ground to aid in the efficiency of the resulting array. The maximum height of each antenna support will be 2 m (about 6 ft). Initial funding will see the construction of 15,000 of these simple, cheap antennas in the Netherlands spread in cluster of spread over 100 km. The prototype with 100 clusters of antennas should be built in 2006–2008. It will then be expanded to across the German border to encompass a total distance of 350 km with 25,000 relatively unobtrusive structures. This enormous radio telescope will have no moving parts.

LOFAR has another unique feature. In addition for radio astronomy studies of the early universe at 10, 100, 150, and 200 MHz (wavelengths of 30, 3, 2, and 1.5 m, respectively), it will be used for studies of crop growth and seismic shifts. The cluster of antennas will be equipped with biosensors and weather stations related to studies of crop growth over an enormous area, given that the radio astronomers are already linking up all those antennas electronically. Why not add some additional, useful, information while you are at it? Similarly, by adding geophones and seismic sensors, data relevant to what's going on under the Dutch and German countryside, believed to be sinking with the removal of natural gas, will also be obtained.

A consortium of European observatories is involved in making this project a reality with the Astronomical Institute in Dwingeloo, the Netherlands, the driving force and with collaboration from groups in Germany, Sweden, France, and the United Kigdom. It is expected to be fully operational by 2015. Given that LOFAR is extending both sensitivity and resolution by 100 to 1,000 times in a largely unexplored frequency range we can expect surprises once it becomes fully operational. The history of astronomy has shown over and over again that totally new discoveries follow on the heels of opening a new part of the radio band, after which the discovery rate drops and the study of details takes over.

## 15.4.3. SKA—The Square Kilometer Array

Not to be confused with the notion of an array, that is, square and one kilometer on a side, this array will have a total collecting area of one square kilometer made up of several thousand small dishes in the 10–15-m diameter range spread over a total area of at least 3,000 km across and capable of operating from 100 MHz to 25 GHz (3 m to 1.2-cm wavelength). About half the collecting area will be concentrated inside a few kilometers to enhance the sensitivity of the array to weaker radio sources and to the radio emission from interstellar HI. To date, $00 million has been invested in the design and two prototype arrays will be built in Australia and South Africa. Both of these nations are vying to locate the final array within their borders. Other nations hungering for the SK are Argentina and China. Clearly the dry high plains of the western part of South Africa would be ideally suited for the SK, which would place South Africa well and truly on the radio astronomical map. (This is an unbiased statement by this South Africa born author!)

The sensitivity of the SK will be 100 times better than the VLA, which was used to make most of the radiographs in the book. The SK will have a resolution of a thousandth of an arcsecond at a frequency of 20 GHz (wavelength 1.5 cm). With an instrument like that radiographs will look as detailed as the best optical images from the Hubble Space Telescope. One of the drivers behind the construction of the SK is the challenge of observing magnetic fields in distant galaxies, as well as our own of course, where it is clear that up till now we have been resolution-limited in being able to see details.

Back in 1997 a consortium of nations (Australia, Canada, Chile, India, the Netherlands, and the United States) signed a memorandum of understanding to explore the concept of the SK and the project is being run from the Astronomical Institute in Dwingeloo, the Netherlands. Its anticipated cost is a staggering $.2 billion. But with great patience and a systematic approach to realizing the full array, the radio astronomers involved may yet bring this project to fruition.

## 15.4.4. PAPER—Portable Array to Probe the Epoch of Ionization

In the beginning was the Big Bang, which for all (un)practical purposes can be thought of as a phase in which all was radiation (not light, but very high energy radiation). As the universe cooled, a time was reached when this radiation gave rise to elemental forms of matter inextricably mixed in with the radiation to the point that the radiation was trapped in the expanding universe. Then there came a time when the matter and radiation were able to uncouple themselves, which set the radiation (then in the form of light) free to escape in all directions as the matter itself also moved apart as space expanded. At this epoch, about 300,000 years after the Big Bang, the universe became transparent to its own radiation. All very bizarre, you might think, and it is! The point that lends itself to an astronomical test is that small density fluctuations existed back then that themselves cooled further to the point that the elemental constituents of matter, specifically electrons and protons, combined to form hydrogen atoms. Some enterprising theorists have claimed that when stars begin to form and shine their ultraviolet light will then cause those hydrogen atoms to undergo the transition discussed in Chapter 7, which then leads to the emission of the hydrogen spectral line.

At first thought one might wonder whether this signal has already been seen, except for a mighty fly in the ointment. We are talking about events in the very early universe, which, from our point of view, happened very, very far away. And given that the universe is expanding, that means the radiation from any object back then, whether a quasar, a radio galaxy, or hydrogen gas, will exhibit a very large redshift. So large in fact that the 21-cm hydrogen line, whose initial frequency is 1420 MHz, will be shifted into the range of 90–190 MHz by the time it gets here. The lower limit is set by the conditions that created the first stars, and the other end of the range is set by a time when there were so many stars that the universe essentially heated up again to the point where the hydrogen gas all became ionized. That is known as the epoch of reionization.

An enormously ambitious radio telescope capable of searching for very weak radio signals in the frequency range 90–190 MHz is now in an advanced stage of development with a 32-element array about to come into operation at the NRAO in Green Bank. Named PAPER, for portable (or precision) array to probe the epoch of reionization, investigators from the University of California, Berkeley, the NRAO, and the University of Virginia, have taken up the challenge of detecting a very faint (1–10 mK)signal in the presence of terrestrial radio interference. After testing of what they call PAPER-32, they will raise their sights by constructing a 128-antenna array (PAPER-128) in the Australian outback. The antennas are relatively cheap to build, and the data handling capability will be very sophisticated. All-in-all this project presents a tremendous technical challenge.

# 16
# What's It All About?

## 16.1. Expecting the Unexpected

The very high resolution radio telescopes now being used or planned will help us understand many of the secrets of the radio universe, but unless new windows into the universe are opened up (and there are a few left for radio astronomers—the submillimeter wavelength region, for example) we may expect the flood of dramatic new discoveries to dry to a trickle. Radio astronomy has matured and to a large extent it has entered an era of consolidation, of making more detailed observations of ever more detailed phenomena.

Looking back on the last 50 years, it is striking that many of the most important breakthroughs have occurred because of accidental discovery (sometimes called "serendipity") while the astronomer was expecting to study some other phenomenon. This often happens in a new science or, in the case of astronomy, when a new part of the spectrum is opened to our gaze. However, it takes more than being at the scene to become a faithful witness to a new phenomenon. It takes openness to the unexpected, as well as willingness and skill to follow the clues nature provides, wherever they may lead. Only then will the scientist be primed to discover something new upon the face of the sky. In the pulsar saga this has happened at least twice.

It is worth repeating that serendipitous, apparently accidental, discoveries have played a key role in the history of radio astronomy, but no one can make a discovery without being prepared for it. As Louis Pasteur once said, "In fields of observation, chance favors only the mind that is prepared." The important discoveries were made because the right person had the right equipment and was doing what turned out to be the right experiment at the right time, even if the results turned out to be unexpected.

Pioneering discovery usually occurs in a climate of "tradition," in the context of accepted "models" or ideas about the way things are. Within the scientific establishment, tradition and authority rule as much as they do in other systems of organized belief. However, the very essence of the scientist's philosophy is built on the awareness that scientific knowledge will change, despite inertia associated with tradition. Consequently, our fund of knowledge about the universe and our interpretations of natural phenomena systematically move toward new levels of understanding; that is, they evolve. The potential for progress through the evolution of ideas is built into scientific methodology. Therefore, when a researcher has that peculiar combination of skill, luck, persistence, and a willingness to risk the censure that may accompany the expression of new ideas, there is a chance that he or she may achieve an exciting breakthrough and contribute measurably to "progress."

## 16.2. Are We Still Open to the Unexpected?

Consider that pulsars and their near twins millisecond pulsars would not have been discovered serendipitously if human intuition and persistence had not been directly involved in the process. As radio astronomy matures, the increased use of supersophisticated computer technology in making observations often removes the human factor to the point where serendipitous discovery becomes well-nigh impossible. A computer can only do precisely what we want it to do, which means that we have to know what we are looking for before we search! Then the computer either finds what we want to find or does not. This implies that important discoveries such as Jocelyn Bell's discovery of "little bits of scruff" on paper chart recordings unsullied by computers might become ever more rare. It is possible that our technological sophistication has evolved to the point where we may no longer be able to see the unexpected.

## 16.3. How Much Longer Will Radio Astronomy Last?

We have traveled the invisible universe of radio astronomy and taxed our imaginations as we struggled to comprehend the enormity of radio galaxies and quasars and the pervasive presence of black holes in space. We have visited clouds of molecules between the stars and seen the wonders at the center of the Galaxy. Now we should stop for a moment and ask how much longer will radio astronomy last?

This question is not asked lightly. Radio astronomers observe in the radio frequency part of the electromagnetic spectrum, but there are many people who would like to use those same radio bands for other purposes. Some of the more obvious groups (or services) with an interest are the communications and entertainment industries, the military, and NASA. Cell phones, wireless networks, GPS, and a host of other "new fangled" uses of the electromagnetic spectrum come at a price. They all need parts of the radio band and pose a potential threat to radio astronomy by generating unwanted radio interference.

A radio signal leaking from a communications satellite can destroy hours of astronomical observations. To the radio astronomer such a signal appears as a flash bulb might appear to an optical astronomer trying to take a photograph of a distant galaxy. The radio frequency spectrum is a natural resource that is increasingly being commandeered by those who wish to use radio frequencies for commercial and military uses. Radio astronomers feel this takeover very keenly and although they have a voice in the World Administrative Radio Conference, which decides on how to share the radio spectrum, their lobbying power is not backed up by the dollars available to commercial and military interests. On the positive side the hydrogen-line band around 1,420 MHz is well protected, as are several other bands centered on the spectral lines of astronomically interesting molecules such as OH.

Basically the radio astronomers want to keep the radio spectrum as quiet as possible. All they want to do is listen to faint cosmic whispers. The irony is that while international agreement has created protected bands for the various services, such "protection from services in other bands shall be afforded the radio astronomy service only to the extent that such services are protected from each other." Most other services do not care how much spurious radio energy is leaking from one radio band into another, because such leakage is usually well below their levels of interest. However, this leakage, due to poor design of transmitters, is a continual threat to radio astronomy. At issue is the fact that while a given service does make some effort to prevent excess leakage of unwanted signals into other bands, this concern is relative. Radio telescopes, capable of picking up faint signals from quasars 13 billion light-years away, are readily swamped by satellites leaking radiation in directions of little concern to their users.

The radio astronomy community has unusual needs, which those interested only in communications do not always see as a priority. The continuing threat is that the wonders of the invisible universe, now being so dramatically revealed, may be rendered invisible because of radio pollution generated by the technological nations on our planet.

## 16.4. So, What's It All About?

We've seen what radio astronomers do and what they have found. But there is another question that I am often asked, as are most scientists at some time in their lives: "Why do you do this? What is the use of radio astronomy?" or words to that effect.

My first reaction used to be puzzlement that something that fascinates one has to be useful to be relevant. Whatever happened to basic curiosity? But then in our country (in the United States), curiosity is not something that is encouraged in vast swaths of society. After all, curiosity is dangerous. If you exercise your curiosity you may discover something that is not in accord with what you believed up to that point in your life and, worse still, you might discover something that is a threat to the beliefs of the group of people amongst whom you were raised or reside. It is the inherent threat posed by unfettered curiosity that causes so many individuals to retreat into their shells where they feel safe in their beliefs, beliefs that are invariably not tested against reality. As an example, without wishing to get into esoteric arguments about the nature of reality, it is worth looking at the debate about evolution. If ever there was a discovery that emerged from the human expression of curiosity about the nature of life, this one takes the cake. Darwin is largely blamed for the idea, but since his epochal work *The Origin of Species* countless other scientists have contributed to the evolution of ideas about the nature of evolution. And yet, 160 years later, there remain large segments of society whose intense adherence to ancient beliefs is profoundly threatened by the concept that human beings exist on this planet as the result of the inexorable workings of natural processes, which we lump together under the all-embracing label of evolution.

While it may appear that I am digressing, the fear of evolution expressed, in particular, by fundamentalists of all colors, epitomizes the question I began with. What is the use of radio astronomy, or for that matter what is the use of most of modern-day scientific research? One answer is that all of modern civilization rests on the findings of scientists who could not have foreseen where there research was headed, but whose ideas were then brilliantly exploited by engineers who invented devices that operate on principles unearthed by fundamental research.

Furthermore, if society as a whole paid closer attention to what scientists have discovered our world would be a very different place. For example, the discoveries of radio astronomy have enormously broadened our view of ourselves in the cosmic scheme of things. Imagine that everyone was aware of the nature of our universe as revealed to scientists. Might that not cause us to rise above the petty rivalries that have for so long motivated human beings to kill each other over beliefs?

Setting such idealism aside, my thesis is that when one becomes aware of the threat to human existence of comets and asteroids, or that star death can wipe out all life on planets within 50 light-years of the conflagration, once we recognize that in distant space entire galaxies are being torn apart by gigantic explosions in their nuclei, once we appreciate the scale of catastrophic events in AGN (active galactic nuclei, an understatement of the true state of affairs, if ever you've seen one), once you recognize the existence of stupendous physical phenomena that play themselves out without paying any attention whatsoever to whether or not there might exist sentient beings on some distant planet capable of perceiving them, then surely we must pause for thought.

The thoughts that surge through me when I think about what I have written in this book, about the chaos at the nucleus of our Galaxy, the existence of microquasars and pulsars, neutron stars and black holes, radio galaxies and quasars, stand in such stark contrast with events portrayed on the evening news that I wonder why it is that we continue to behave as if our continued existence on this planet is assured forever. Nothing could be further from the truth.

There exists a sect on the fringes of radio astronomy that peddles the bizarre notion that if only we could detect radio signals from distant civilizations we would learn the secrets we need to hear, secrets to aging and the curing of diseases, for example, or how to assure the survival of civilization. Yes, believe it or not, that is a primary motivation for the search for extraterrestrial civilizations, driven by a generation raised on science fiction myths that would have the Galaxy populated by sentient being just aching to talk with us. Of course, extraterrestrials may be aching to convert us to their beliefs because why else would they be broadcasting for us to hear? A parallel exists on our planet. On the short-wave radio bands these days, the most powerful transmissions are generated by Christian missionary sects wishing to spread their word to other continents.

The point is that the usefulness of radio astronomy, or of any of the basic natural sciences, is barely recognized in part because of inadequate schooling, but also because it is very difficult to keep the public informed about the staggering and extremely rapid progress being made in labs and observatories around the world. At the same time, if we were to fully communicate what has been learned our story might profoundly upset those whose worldviews are shaped by ancient beliefs. I would go so far as to suggest that the reaction to what Darwin hypothesized is nothing compared to the reaction that those same segments of society would experience if they fully understood what scientists the world over have discovered about the nature of life and the nature of our Universe. There is a hint of things to come in the Christian fundamentalist's fear of the notion of the Big Bang. That is only the beginning.

For me the relevance of radio astronomy is that it helps us perceive our place in space. It gives us a much-needed perspective on who we are, as seen in the context of the grander scheme of things. What we find is not encouraging. We exist in a universe of awesome and incomprehensible dimensions in space and time. To some extent all of science adds to this perspective, and yet we continue to behave as if this planet will be here forever, that our civilization will exist eternally, and therefore that we need not pay any heed to the atrocities we foist upon one another. So my dream is that when we begin to fully appreciate the awesome magnificence of the physical universe in which we find ourselves, in which we are now capable of peering to the very edge of space and time, that we will take note and realize that in the mindless play of nature out there we are barely even pawns. Therefore, why would we act to hasten the demise of our species by prejudice, intolerance, wars, or terrorism? Isn't it time to really pay attention to what scientists are doing and what they are finding, and then let us reflect on their discoveries and hypotheses and mull that over, without fear that our sacred beliefs will be threatened by so doing. They may well be, for that is the nature of evolution—the process of replacing the old with the new—but to avoid this confrontation is to avoid the facts of our existence.

Once we begin to confront the full, awesome scale of the universe in all its detail, perhaps then we can exercise our minds and our imaginations in the manner to which we have evolved as a thinking, intelligent species on a very, very tiny planet in a corner of a very, very large Galaxy in a Universe that has existed for more than 13 billion years.

# Appendix

## A.1. "Seeing" Radio Waves

Radio astronomers talk about "seeing" radio sources. This is a figure of speech, because they do not literally "see" radio waves, nor do they even "listen to sounds" from space. Cosmic radio whispers are far too faint for the human ear to perceive, even after the radio signals have been amplified a million times. Instead, modern computer graphics technology allows the radio waves to be converted into electrical signals, which can be combined to produce an image of what the radio source would look like if you had radiosensitive eyes. Such images are sometimes called radiographs, which are found throughout this book. All forms of data produced by radio telescopes, whether as radiographs or just sets of numbers, are examined visually on a computer screen or in hard copy and thus radio astronomers talk as if they can "see" the objects of their study.

It has become fashionable because of its dramatic impact to add color to radiographs where the color corresponds to a gradient of intensity of the received signals. However, I have converted many of the available colorized images to a gray-scale representation because in most cases the eye can pick out more detail in such a display. Also, it helped keep down the price of this book! I trust the original investigators will forgive me for not using their excellent colorized images.

## A.2. The Electromagnetic Spectrum

Today astronomy is far more than peering through optical telescopes. There is radio astronomy, Xray astronomy, IR astronomy, UV astronomy, and even gamma ray astronomy, each with its special types of telescopes and each sensitive to an invisible universe different from the one we can see with our eyes.

All these radiations are part of the electromagnetic spectrum. All electromagnetic waves (so named because they contain an electrical and a magnetic aspect) travel through space at a speed of light 300,000 km/s.

| Frequency | Wavelength |
|-----------|-----------|
| 10 MHz | 3,000 cm or 30 m (shortwave band) |
| 100 MHz | 300 cm (FM band) |
| 1,000 MHz | 30 cm |
| 1,420 MHz | 21 cm: hydrogen line] |
| 10,000 MHz or 10 GHz | 3 cm |
| 100 GHz | 3 mm |
| 1,000 GHz | 0.3 mm (submillimeter band) |

## A.2.1. Wavelength and Frequency

Electromagnetic waves have a wavelength and a frequency which are related in a simple manner. Consider waves crashing into the seashore or lapping the edge of a lake. You may notice that the waves strike the shore at a certain rate, or frequency. Along the California coast a typical interval between waves is 10 s, which indicates a frequency of 6 waves per minute. The frequency at which the water waves break is related to their speed. The faster they travel the greater their frequency. There is also a characteristic wavelength—the distance between the crests of these waves.

As regards radio waves, we use frequency in the text and the conversion to wavelength is straightforward, as shown in the table. Frequency is measured in cycles per second called a Hertz (Hz), after the physicist Heinrich Hertz. A million Hz is written as MHz. A billon Hz is a gigaHertz or GHz.

A simple way to convert from wavelength to frequency or vice versa is to remember that wavelength times frequency is equal to the speed of light.

## A.2.2. The Wavelength Range of the Electromagnetic Spectrum

Radio waves are at the long-wavelength end of the electromagnetic spectrum, as long as hundreds of meters down to a millimeter and well-known "microwaves" have wavelengths of around a few centimeters. As wavelength decreases we find infrared (IR) radiation, commonly experienced as heat, then light waves which range from 70 millionths of a centimeter ($7 \times 10^{-5}$ cm) for the long wavelength red light down to 40 millionths of a centimeter ($4 \times 10^{-5}$ cm) for violet light. The colors of the rainbow fall between these two extremes. Beyond that going to shorter wavelengths comes ultraviolet (UV) radiation, sometimes called "blacklight." (UV causes sunburn, and in large doses is extremely harmful to living organisms.) Next are Xays, with wavelengths so short that they literally wriggle between atoms and so can penetrate our bodies. Finally, at the shortest (less than $1 \times 10^{-8}$ cm) wavelength end of the spectrum are the gamma rays.

## A.2.3. Atmospheric Windows

The earth's atmosphere cuts out most of the UV, IR, Xays, and gamma rays from space. However, the atmosphere is transparent to radio, light, and some infrared waves, providing two "windows" into space. Radio astronomy can usually be done during the day or night, independent of the weather. However, water vapor ($H_2O$) and carbon dioxide ($CO_2$) in the atmosphere absorb the shortest wavelength cosmic radio waves. Terrestrial clouds are in fact opaque to millimeter waves, while 1-cm radio waves are partially absorbed. Short-wavelength radio astronomy can best be done in dry climates at high altitudes (in order to be above as much atmospheric water vapor as possible). IR, UV, X-ray, and gamma-ray astronomers generally require either balloon-borne telescopes or spacecraft in order to make their observations above the earth's protective blanket because the nitrogen and oxygen in the atmosphere absorb these radiations.

## A.2.4. Spectral Lines

Light from distant stars, galaxies, and quasars often contains energy at very specific wavelengths. These signals are known as spectral lines. Suffice it to say that spectral lines are usually due to radiation of energy from particular atoms or molecules and that each has its own characteristic signature of spectral lines which astronomers are trained to recognize.

## A.2.5. The Redshift and the Doppler Effect

The redshift is the name given to the stretching of light waves, or any other electromagnetic waves, produced by movement of the source of the radiation away from the observer. To illustrate this, recall the sound of a jet plane. When it is flying toward you the sound is high-pitched. As the plane passes and flies off into the distance the sound moves to a lower and lower pitch. This is known as the Doppler effect, after the Austrian physicist (Christian Doppler) who first studied the phenomenon of the change in frequency of waves from a moving source. When the plane flies away the sound waves it emits are stretched—that is, their wavelength appears greater. That produces a lower tone. Similarly, when a galaxy or star is receding from the earth its light waves are slightly stretched, which means they shift toward the longer, or red end, of the spectrum—hence a redshift. The opposite effect, produced by the object coming toward you, would produce a shortening of the waves, or a blueshift.

## A.2.6. Velocities in Radio Astronomy

In radio astronomy the velocities for galactic hydrogen (Chapter 6) are given in terms of a Doppler shift of the 1420 MHz spectral line with respect to a reference known as the local standard of rest (lsr), defined by international agreement. The sun actually moves with respect to the lsr, which is representative of the way local stars and gas are moving as a whole through the Galaxy. It is useful to refer to velocities of distant gas with respect to this standard because the earth is constantly moving through space and at any given moment the observed Doppler shift of a distant hydrogen cloud, for example, depends on how the earth is moving with respect to it (or vice versa) in its orbit about the sun. This convention is different from the one used by optical astronomers who define an object's (star or galaxy) velocity with respect to the Sun as the reference point.

## A.3.  The Brightness of Radio Sources

The strength, or intensity, of radio waves received from a distant radio source is usually given in terms of units defined by international agreement. Such a unit is the Jansky, named after Krl Jansky, the discoverer of the radio waves from the Milky Way. A Jansky is a measure of the amount of radio energy striking a given area (one square meter) in a specific frequency interval (1 Hz) and is equivalent to $10^{-26}$ watts per square meter per Hertz, not something that rolls readily off the tongue. Radio astronomers also describe the intensity or strength of a received radio signal in terms of a temperature. The antenna temperature, given in terms of degree Kelvin, is the temperature the universe would be at in order to radiate the same power as is captured by the radio telescope observing that specific source. Luminous radio sources may produce very large antenna temperatures, depending on whether or not they fill the beam of the antenna. For a very small, point-like radio source, which may be intrinsically bright, a small antenna temperature will be produced because the source covers a small area of sky as compared with the beam of the telescope.

The luminosity of a radio source refers to the actual amount of energy it emits whereas the brightness of the source is a measure of the power per unit area radiated by the source. For example, a flashlight may appear bright when placed close to one's face, but across a football field it will appear quite faint. Distant stars also appear very faint in the night sky, but if we should move close to a star we would find that it is enormously more luminous than the flashlight. These terms can be used to refer to optical or radio emission from astronomical objects.

For completeness, it should be pointed out that the measured brightness (or brightness temperature) of a radio source can be calculated from the observed antenna temperature if the source diameter and the beamwidth are known. The brightness temperature can then be used to infer the amount of energy actually generated at the radio emitter, provided the distance of the source is known. That, in turn, allows the physics of the source to be better understood.

## A.4.  Radio Spectra—Identifying the Emission Mechanism

In order to tell whether the radio source is thermal or nonthermal the radio astronomer measures the intensity of the received radio waves at many widely separated wavelengths. The spectrum of the radio source is defined as the manner in which the intensity of the received radio emission varies with wavelength. The spectrum of a synchrotron source shows that it becomes brighter at longer wavelengths. The brightness of a thermal source, on the other hand, decreases with increasing wavelength. Determination of the brightness of a radio source at several radio wavelengths allows its spectrum to be determined, and this is usually enough to show whether the radio source is thermal or nonthermal in nature. This, in turn, allows the physical conditions in the source, such as temperature, density, and magnetic field strength, to be determined. Nonthermal sources include quasars, radio galaxies, and supernova remnants. Thermal sources include the sun and clouds of hot gas (HII regions) that surround young stars. In addition thermal sources are unpolarized and nonthermal emitters may show significant degrees of polarization.

## A.5. Notation

One million can be written as $10^6 = 1,000,000$, i.e., a 1 followed by six zeros. For numbers smaller than unity the notation is similar, e.g., $10^{-2} = 0.01$, or one hundredth. A light year is about $6 \times 10^{12}$ miles. More useful to remember is that a light-year is nearly exactly $10^{18}$ cm, or $10^{13}$ km.

This superscript notation is also used in another way. If a gas has a density of a million atoms per cubic centimeter, astronomers write that as $10^6$ cm$^{-3}$.

A common tradition in astronomy is to refer to the mass of astronomical objects in terms of the mass of the sun. One solar mass is about $2 \times 10^{33}$ grams, a number far too great for us to comprehend. It is easier to describe star's mass in terms of solar mass, which gives us something to sink our imaginations into because now we can relate astronomical mass to something closer to home.

## A.6. Position Measurement and Angular Accuracy

In order to identify the source of the radio waves, radio astronomers need to determine an accurate radio source position so that they can compare these with the optical photograph of the same part of the sky.

Positional accuracies are usually measured in arcseconds. To appreciate the smallness of an arcsecond imagine looking at your fingernail from two miles away. The fraction of your panorama filled by the fingernail is then about one arcsecond. The largest radio telescope system, the Very Large Array in New Mexico, can discern detail down to about a tenth of an arcsecond.

Note: Angles are measured in degrees ($^\circ$), minutes ($'$), and seconds ($''$) of arc. A full circle contains 360 degrees. Each degree consists of 60 arcminutes, and each arcminute consists of 60 arcseconds. Our fingernail would be about 20 miles away for it to cover an angle of one tenth of an arcsecond.

The ability to discern detailed structure depends on the beamwidth, or the resolution, of the radio telescope. The resolution can be calculated by dividing the diameter of the reflector by the observing wavelength. Thus the 100-m (330-ft) diameter Robert C. Byrd Green Bank Telescope of the National Radio Astronomy

Observatory (NRAO) operating at 20-cm wavelength has a 9-arcminute beam, which means it can "see" details down to 9 arcminutes across. Anything smaller will be blurred by the resolution of the dish and will appear to be 9 arcminutes in diameter. For comparison, the human eye cannot distinguish anything smaller than 20 arcseconds across. This limit is determined by the wavelength of light (about $5 \times 10^{-5}$ cm) divided by the diameter of the pupil (about half a centimeter). In practice the lens is not perfect and sets a limit on our capacity to see details to about 1 arcminute.

Aperture synthesis is the technique of combining signals from an array of radio telescopes spread over large areas of countryside, or even across half the earth, so as to obtain the resolving power of a single dish whose diameter would have to be hundreds or even thousands of miles, an utter impossibility. The largest aperture synthesis telescope is the Very Long Baseline Array referred to in Chapter 15, which can achieve a resolution measured in millionths of an arcsecond.

## A.7. Astronomical Coordinate Systems

The geographical coordinates of latitude (north–south) and longitude (east–west) are used to locate objects on the surface of our planet. Astronomers use similar angular coordinates to locate objects on the sky.

Imagine drawing a line across the heavens that is always directly above the earth's equator. This is known as the celestial equator. The angle measured north and south from this celestial equator is called declination (directly equivalent to latitude on earth). The North Celestial Pole, located directly over the North Pole of the earth (just about where the Pole Star is found), is at $+90°$ declination.

The astronomical equivalent of terrestrial longitude, the coordinate measured east and west of Greenwich, England, is called right ascension, and can also be given as an angle around a circle, but is commonly measured as a time—24 h span the equator, equivalent to a full circle of 360 degrees. Right ascension, in hours, minutes, and seconds of time, is measured east of an agreed upon zero-point known as the First Point in Aries. Although there are technical complications associated with the precise definition of these coordinate systems due to the precessional wobble of the earth over a 26,000-year cycles, suffice it to say that right ascension and declination are the basic astronomical coordinates.

Another system of coordinates is based on the Galaxy itself. A line defining the central plane of the galaxy, a line that runs along the center of the Milky Way band of stars, is defined as the galactic equator. Galactic latitude (b) is measured in degrees and minutes of arc north or south of this equator, and galactic longitude (l) is measured in degrees and minutes along the galactic equator, using the direction of the center of the galaxy, in the constellation of Sagittarius (see Chapter 5), as the zero point.

# A.8.  Astronomical Distances—Looking Back in Time

Astronomers cannot avoid seeing back in time when they look out into space. Light and radio waves traveling at the speed of light come to us from great distances and have been on their journeys for long periods of time. Astronomers are doomed to peering into the past! They are used to this concept, and to them it is second nature to think of great distances in terms of vast spans of time. The most often used unit of distance, the light-year, is based on the distance a light beam can travel in a year. The term "light year" allows us to encompass a huge distance ($6 \times 10^{12}$ miles or $10^{13}$ km) in two words. Astronomers prefer a more rigorous unit, the parsec, which is the distance at which the sun–earth distance appears to cover an angle of the sky of one arcsecond. In astronomical parlance, it is the distance at which the sun–earth line has a *par*allax of one arc*sec*ond, hence parsec.

# A.9.  Keping Things (Radio) Quiet

Radio astronomy is a total mystery to most people, which is why the large radio telescopes such as the Robert C. Byrd Green Bank Telescope, the Australia Telescope, the Bonn radio telescope, and the giant dish near Arecibo, Puerto Rico, exert such a powerful influence on the human imagination. At the Arecibo Observatory their new science center draws 120,000 visitors a year who get a close-up view of the 1,000-ft diameter dish carved into a valley between rolling hills. At the NRAO site in Green Bank, WV, among lovely rolling hills, the visitor can enjoy a bus tour of the site and wander about in the delightful visitor center and its small theater showing a movie about radio astronomy.

The world's largest radio telescopes are located far from cities that generate staggering amounts of radio interference, which inevitably ruin observations of distant radio sources. The NRAO is located inside the National Radio Quiet Zone where no transmission of any radio signals is allowed for dozens of miles in all directions. As a result you will find that your cell phones will not work there. In addition, the receiving equipment for the new giant telescope is installed in an electrically shielded room to block out any remaining spurious transmissions, from passing portable computers, for example.

Thanks to the continuing efforts of the observatory staff, and at similar facilities around the world such as the huge dish at Arecibo, radio astronomers can still tune into the faint radio whispers from the cosmos.

# Index